Construction project management
Getting it right first time

John F. Woodward

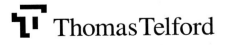

Thomas Telford

Published by Thomas Telford Publishing, Thomas Telford Services Ltd, 1 Heron Quay, London E14 4JD

First published 1997

Distributors for Thomas Telford books are
USA: American Society of Civil Engineers, Publications Sales Department, 345 East 47th Street, New York, NY 10017-2398
Japan: Maruzen Co. Ltd, Book Department, 3–10 Nihonbashi 2-chome, Chuo-ku, Tokyo 103
Australia: DA Books and Journals, 648 Whitehorse Road, Mitcham 3132, Victoria

1139605 9

Learning Resources
Centre

A catalogue record for this book is available from the British Library
ISBN: 0 7277 2557 2

Typeset by MHL Typesetting Ltd, Coventry
Printed in Great Britain by Bookcraft (Bath) Limited

To
my wife and family,
all of whom contributed to this project.

Foreword

The modern applied science of Project Management has evolved over the past twenty five years and is now mature. It now has its own professional institutions around the world. It is widely taught. It is the subject of respectable academic research. Above all, those who practice the science can relate cause to effect with reasonable reliability. Projects can be managed in such a way that more demanding targets for success can be set than formerly — with a high probability of achieving them.

John Woodward has been an active player in the field throughout the evolutionary period of modern project management. He has played an important part in the development of the institutions. He has taught the subject widely. He has carried out and directed research. He has been involved in the real management of real projects. He has achieved an integration of these functions in his own varied professional activity that is a tribute to his intellect and humanity.

It is appropriate and timely that he should now have distilled his distinctive and comprehensive knowledge of the theory and practice of project management in this book. Its breadth and depth commend it to everybody who needs to refresh or extend their knowledge of the subject and it is an ideal text for students. It covers the full range of technique from the hard basics of network analysis to the softer but vital human aspects. It centres upon management of construction projects but draws in and illuminates the points where the techniques are applicable to projects of all types. It contains full references to and commentary upon the work of other writers. These will help readers to find their way around the hinterland of all or any of the particular areas of project management which they need to explore.

As John emphasises in Chapter 13, uncertainty is in the very nature of project management. World class project management is all about shrinking the besetting uncertainties and bringing the work in each project under control. 'Getting it right first time' is a very tough target in this environment. The probability of doing so is high for any project manager who absorbs fully what is set out in this book.

Dr Martin Barnes

Preface

The purpose of this book is to convey the essentials of project management and the principal features of its approach, and to offer these in a book suitable as a text on courses, e.g. postgraduate courses in project management, short industrial courses, specialist modules in Master of Business Administration programmes, and similar courses in construction or general management. The basic treatment of the approach makes the book appropriate for students of project management wherever this occurs in undergraduate and similar programmes, and also for those who wish to conduct their own self-taught studies. The emphasis of applications is in all sectors of the construction industry, but the relevance of project management to any situation of change management in any undertaking is made clear, as is indicated in the listing 'Areas of application' later in this preface. This shows that the methods can be applied to 'change management' in any industry. The book seeks to convey the essential approaches and methods of project management that make it a significant and distinct discipline.

THE STRUCTURE AND STYLE OF THE BOOK

The first three chapters of the book are largely discursive and seek to set the project scene, show why project management has become a subject in its own right and indicate how it differs from other forms of management. While some readers may feel that this is a little simplistic in parts it is designed to generate the feeling for a 'project management' approach and the operating environment in which it is applied. For those who may be tempted to ask the question 'What is the relevance of this to the practice of project management?' the answer is to say that all will be revealed in good time; it is felt to be necessary to give some coverage to the general subject of management, so that the reader may see clearly the ways in which the management of projects differs from the management of other undertakings, and of course the ways in which it has similarities. It also serves to put forward what is perhaps the underlying theme of the whole book, namely that project management is not simply a bag of tools, as was at first thought, but centres upon a broad approach, perhaps even a philosophy, and does require a certain attitude of mind. Readers who are less familiar with the project-based industries will find that the first three chapters give a familiarisation with the world of projects, aided by reference to a number of cases, some real, some fictional, and they will then have some feeling for the 'project management approach'. Perhaps only a limited number of people can ever fully develop this attitude and become really great project managers, in the same way that there are few great engineers or architects; but many more will have to work along with project managers in a project management context. It is therefore important that all who work on projects have an understanding of how these can and should be managed, and what the essential features of the approach are.

The book sets out to be a 'how-to' book and gets down to the nuts and bolts from Chapter 4 onwards. This part of the book gives considerable detail about the most useful techniques, backed up by several examples, mostly drawn from the author's experience in the management of projects. For those readers who recognise some of the cases it should be said that they have mostly been modified either for simplification, to hide the identity of participants, or to make the arithmetic less arduous. The text is also based on many years experience of teaching project management to a variety of groups in different industries; this has prompted emphasis to be placed on those topics which some students have found difficult.

PUBLISHING CONTEXT

At the end of the book there is an index of subjects covered plus a bibliography of some of the books published in the area of project management. Where possible brief notes about each book are appended to give readers an opportunity to decide whether a particular book is relevant to their needs and interests; it is in no way intended as a critical review of each book listed. Some of these books are now out of print and/or out of date, some are aimed at very specific types of project, others concentrate on the history and evolution of project management; a few of these give an excellent background to the subject and will probably not become obsolete. There are also many books which are based on academic work and make a very valuable contribution to our theoretical understanding of the subject; they are sometimes a little too theoretical to be used for teaching the basics of the discipline, or for use by individual readers who are seeking a basic, practical introduction.

MANAGEMENT AS A SUBJECT TO BE STUDIED

The study of management has expanded enormously over the second half of the twentieth century; indeed business and management studies together form probably the most popular subject group in colleges and universities. Whether this is due to an explosion of knowledge or to the perception of business as a good career may not be clear, but there is no doubt that the careful study of management has become an academic discipline which offers enormous benefit to the conduct of undertakings. The early years of the analysis of management were largely devoted to the mass-production industries through operations and production management with their attendant techniques of quality control, work study, inventory control, and later, operational research. In parallel were studies related to the structure of organisations, dealing with people, financial control and other techniques which are applicable to a wide range of undertakings. These studies and many others are now well established and continue to develop. The range of management literature is vast, as can be seen from the large number of publications in management libraries and bookshops. It is often difficult to select from among these, since titles are not always descriptive of the subject content. Choosing the subtitle 'Getting it right first time' seeks to convey the essential theme of the subject, namely that in nearly all projects the work is done once only; hence the current quality concept of 'getting it right first time' is especially apt. Another connotation of this phrase is that thoroughness is implied, and thoroughness is at the heart of project management. It can therefore be argued that 'getting it right first time by being thorough in approach' is the real essence of the subject. In fact 'Essence of Project Management' was suggested, but this was felt not to be sufficiently distinctive, since this book seeks to encapsulate the characteristic features of project management which differentiate it from other approaches to management. It also presents those approaches, methods and techniques which are of crucial importance to the management of projects, and thereby provides a means of learning about their management in the real world.

Project management compared with operations management

Project management is not simply a modern name for a subject which has been in existence for some time; it is distinct from operations or production management from which it is a new departure. There are several important differences and these form the basis of Chapter 2. These differences largely stem from the fact that projects are one-off undertakings using resources specifically gathered for that project, and on the instruction of a specific client. By comparison operations management is usually concerned with the control of an existing production facility which produces significant quantities of a small number of products in some form of steady-state process, using fixed resources, and offering the products for sale after manufacture.

The history of project management

This topic has been very well discussed by Peter Morris in two of his books (1987, 1994), and these make good general reading; they illustrate by example how the subject has evolved, largely from within the construction, defence and the computer industries. These all have the characteristics of one-off undertakings, albeit that micro-computer manufacture now comes within the realms of mass production. (It is the design and installation of computer systems which is amenable to the project management approach.) There have clearly been very large projects for a very long time, many centuries before anyone had heard of project management as a subject. The management of projects such as the ancient pyramids of Egypt has been frequently discussed, as has that of large civil engineering projects like the Suez Canal. In their own way these were controlled by a type of project management in the form of one very senior person in charge of everything, e.g. Thomas Telford in the case of UK canal engineering. This approach was lost in the first 60 or so years of the twentieth century, but has now been 'reinvented' and incorporated within the discipline of project management. The main point here is that projects have always been managed, sometimes well and sometimes not so well. The development of project management as a discipline therefore brings together the original practice of central planning and control and the development of modern management ideas and techniques, bound together by an ethos which can best be described as the 'project approach'. It should be noted that sometimes reference is made to project management as if it were a fully determined method of managing projects offering a magic solution; this will usually be a vain hope. The discipline of project management is a generally applicable approach to the management of projects, just as personnel management is concerned with nearly all aspects of the management of people.

Areas of application

As stated earlier, this book is largely concerned with applications in construction, but it is important to note that all of the principles and most of the techniques discussed here are relevant to virtually all other industries where a change has to be implemented and managed. It is only in the detail of application in the context of the working practices of each industry that special reference has to be made. Chapters 9 and 11 covers this aspect further; it is sufficient here to note that typical examples would include: the launch of a new product; entering a new market; setting up a new production line; implementing a new corporate policy; opening a new branch office; introducing a revised accounting system; organising a sports or charity event; recruiting and training staff for a new venture; the design of a new product and its testing; a company take-over or re-structuring, or preparing and introducing new legislation.

The major topics of project management

While an idea of these can be seen by looking at the list of chapter headings, the range of topics covered in various parts of the book are briefly summarised below. The index gives a means of locating discussion of these and many other topics within the chapters.

- What is a project, what are its distinctive characteristics? The definition of the content or scope of a project.
- The basic concepts of managing any undertaking, and how these manifest themselves in projects.
- The similarities and differences between project and operations or production management.
- Questioning the mythology that 'if you can manage one type of organisation you can manage another'.
- The broad areas of projects which have to be managed, including scope, organisational structure and contract form, the project manager's terms of

reference, planning, establishment and achievement of quality standards, the control of time and cost, resources, risks, finance and people.

- The basic techniques and methods of project management, including planning and control methods.
- Project appraisal, project optimisation and consideration of long-term costs and liabilities.
- Project management information systems (PMIS).
- Dealing with certainty and uncertainty.
- The implementation of the management of quality in projects in the context of 'quality assurance schemes', e.g. ISO 9000.
- The specific case of the building industry and the roles of client, architect, quantity surveyor, engineer, other specialists, main contractor, sub-contractor, and the project manager.
- The project manager as an individual, the personal characteristics, training and experience needed. The importance of team-building.

Note Throughout this book it is intended that the text is never gender-specific; the term 'project manager' is now in general use for both women and men. It is however difficult to avoid such words as 'he, she, his, her', and the alternatives of using 'he/she, him/her, or s/he' are not yet widely adopted. Similarly the term 'man-hour' has been retained since it is in widespread use and is well understood to have a specific meaning. The alternatives of 'person-hour' or even 'operator-hour' could be construed to have a different meaning; the term 'worker-hour' has been used in a few places.

The author has four daughters all of whom are engaged in the construction industry one way or another, and he is therefore well aware both of the part played by women in that industry, and their wish to be properly recognised. It is to be hoped that this recognition and the words to go with it will soon be generally in place.

Acknowledgements

Many thanks are due to all who have helped me form my ideas of project management. Perhaps firstly to my colleagues in The Taylor Woodrow Group with whom I worked on many major projects, in particular John Ballinger who by his example instilled in me the idea of being thorough and trying to 'get it right first time', which is at the heart of good project management. This echoed the advice long ago from my father who adopted the adage 'If a thing is worth doing it is worth doing well'. These early concepts were developed further by my academic colleagues at the University of Stirling, especially Frank Bradbury, my head of department. Many of my contemporaries helped to develop ideas of project management through the APM; Martin Barnes, Steen Lichtenberg, Peter Morris and many others. I was also helped by colleagues at the University of Paisley where projects took on a wider meaning beyond construction. In recent years my work has extended into an association with architects, partly through my family but in particular with Nick Charlton-Smith of The Royal Incorporation of Architects in Scotland, who has led me to look at ways in which that profession can regain some of its former ground in the management of projects.

In the preparatory work for this book I am indebted to The Leverhulme Trust who supported some of the work, and also to Trinity Hall, Cambridge, where I spent a year as a visiting fellow. I am also very grateful to Professor K. Smith of the Massachusetts Institute of Technology who organised for me to visit and meet many of the great names of project management. In the technical production of the book I have been greatly assisted by Sally Smith and her colleagues at Thomas Telford Publishing, and by my daughter Susan Moor who prepared most of the diagrams. The words and numbers are mine, mistakes and all!

John F. Woodward
Killearn, Scotland

Contents

CONTENTS

1   What is a project?

Readers will probably have a very good idea what they mean by a project, and the context in which it exists, but unfortunately the word has gained widespread usage in a large range of areas which are not really within the scope of our discussion of the subject of project management. For example most school children will be familiar with the task of undertaking a 'project' as part of their studies, whereas their parents would have had to write an 'essay'; the difference being that a project implies a little more research and probably involves the production of diagrams or pictures. Other uses of the word 'project' are common in local authority jargon where a specialist unit is set up to carry out a particular function such as home care for the elderly; but this is perhaps better described as a 'programme' in modern terminology. A general distinction between a project and a programme is that the former usually has a defined life with a clear beginning and end, while a programme has a continuing existence until it is brought to an end. This introduction to the concept of a project is deliberately simplistic in order to discuss the characteristics of projects and hence the ways in which they can be managed. The principal features of a project — as the word is used here and which are of great relevance to project management — are discussed below, but can be summarised in the following list. Note that this list is indicative only and is not intended as a comprehensive definition.

- Each project has the specific objective of creating some new entity which did not exist before, i.e. projects are goal-oriented.
- Projects have a defined beginning and finish, with a clear project life cycle.
- Projects are made up of a large number of separate but interdependent tasks.
- Projects are unique.
- Project tasks make demands on a range of resources, usually on an intermittent or varying basis.

THE PROJECT CONTEXT

It can be seen that the above characteristics could apply in a wide range of different industrial and commercial contexts, and it is worth considering these briefly before entering a more detailed discussion of the above list. Project management as we know it today had its origins in the construction industry and has become widely used there over the past 30 years. Clearly projects have been managed one way or another for much longer than this, but it is only recently that the approach has been rigorously studied and applied. It is important to point out that some people may promote the concept of Project Management (with capital initial letters) as a particular system with specific techniques, but the generality of project management (with lower case initial letters) is that it comprises a wide range of methods and ideas which have been brought together since they are particularly relevant to the management of one-off undertakings, such as building projects, and are somewhat distinct from the package of methods which have developed over a longer period for the management of steady-state operations, such as manufacturing.

What has recently evolved is that many situations in manufacturing are one-off, e.g. a new product launch, and in these cases the methods of project management have something to offer. This book largely concentrates on consideration of the management of projects from the point of view of the building professional, whether client or contractor, architect, surveyor, planner or engineer. This does not mean that the contents are not relevant for other managers of projects; most of the techniques are readily applicable in other industries where any situation of change has to be planned and managed, and these are more fully discussed in some of the cases quoted later. There are clearly a few chapters which are of specific relevance to construction and several, but not all, of the examples cited relate to that industry. Much of the development of project management as a discipline has taken place in construction, but equally important have been the defence and computing industries, where a great deal of the initiative has been taken. The benefits of the methods have been realised, and it is now recognised that very nearly every organisation is involved at some time with organising a project, and that it can gainfully use the project management approach.

The project 'context' as referred to in this book refers to the specific organisational structure of a contract and the parties thereto, its relationship with legal and financial constraints, and the particular people or organisations involved. It is therefore concerned with the immediate surroundings of the project, and will differ from one project to another, even though they may be in similar locations at the same time. It should be noted that there is no general agreement on the use of the terms 'project context' and 'project environment' among writers on the subject. One important authority on project management, Professor J.R. Turner in *The Handbook of Project-Based Management* (1993), makes a different interpretation from that given here; he regards 'environment' as a physical concept, i.e. the site in the case of a building. His view of context is more abstract, and encompasses 'the complete economic, human social and ecosystem of the project'. In his definition therefore the context includes the environment, whereas in this book the reverse is the case. This apparent contradiction need not worry the reader, as the specific definitions are not widely used, but it is important to think about and understand that all the aspects of context and environment, no matter which of the two definitions is used, play an important part in the management of projects.

THE PROJECT ENVIRONMENT

The actual organisation (or 'context') of the project management function in any specific project will depend to a great extent upon the project environment pertaining to that case. In the case of construction it is not only the physical environment of location, ground and weather conditions that apply, although these are of very significant importance (OECD, 1979); the economic environment at the time of the project will also be highly relevant, as the keenness of competition for work varies from time to time, as do the availability of capital and grants. The legislative environment is also important to construction, with planning regulations, conservation areas and listed buildings status. By contrast a project in the computer industry will have a somewhat different environment, and will for example be largely independent of physical location and weather conditions. A computer installation will be much more vulnerable to the rapid development of technology in that industry, where every project is largely obsolete as soon as it starts; this imposes a totally different aspect to the project design, one which is not often present in construction projects. Yet another project environment is that in the petrochemical industry, one which is typified by the high capital investment in fixed plant. In addition to the major projects of new plant construction there are often many smaller projects concerned with the modification and/or maintenance of plant. Here one of the main environmental factors is the urgent need to get the plant back into operation, so that great emphasis is placed on the time control of projects, and cost becomes of lesser importance. Due to the need for high reliability it is always necessary to plan for and achieve high quality standards, again without squeezing costs. There are also particular requirements of safety in regard to the chemical industry, due largely to the potentially widespread and catastrophic nature of a major failure. From these and other examples it can be seen that the project environment and the way in which it changes from one industry to another is very significant for the management of each project.

PROJECTS AS SYSTEMS

The development of project management as a discipline has taken place alongside the systems approach to management (Walker, 1984). For those with an understanding of systems with its definitions of inputs and outputs, its boundaries and controls, it will be easy to see that project management can be viewed in this way; it is not necessary to have such prior knowledge to understand this book, but it will be the case that anyone who gains a good understanding of project management will at the same time almost unwittingly learn something about systems. Many of the more academic books with an analytical view of the management of projects quite rightly use a systems approach to their subject, but for a practical working project manager the understanding of systems theory may be interesting but not essential. Having seen the importance of context and environment and the link with systems thinking let us turn now to a more detailed look at the list of the characteristics of projects noted above.

Project objective

Projects are often said to be 'goal-oriented', which means that most projects will have one main objective. This may be the building of a school or hospital, the acquisition of a new piece of machinery, the introduction of a computer system, submission of an architectural design bid, the elimination of a persistent fault in a production line, the change of use of an existing building, and so on; the list is infinite. Note that in most cases the objective is very simply worded and that it states the desired end result. The first example 'to build a new school' does not list within its objective the design of the school, as this will simply be one of the major components of the whole project, as will the site purchase, planning consent, funding approval etc. The stated objective will be the single essential outcome, and is not intended to be a description of the project as a whole, or even of its leading features. Possible confusion can arise however because different groups of people will have separate subsidiary objectives within the same project, e.g. the job architect may be given the task of developing a brief with the client and then preparing a design to meet that brief; this will then become the architect's own project, which will form part of the overall project of building the school. At the same time, after appointment, the contractor's objective will be to complete the construction at a profit; this must be done in accordance with the terms of the contract and this fact should not need to be stated as part of the objective. It is worth reflecting upon this idea of a single primary objective; the contractor may say that completion on time is important but this is usually both one of the contract terms and also is normally in the contractor's own interest; it is therefore not a specified part of the stated objective.

This concept of a primary desire for profit should be recognised as being very important; it is not a matter of greed on the part of the contractor. One of the worst things that can happen on a project is for the contractor to be forced into liquidation as this will invariably have an adverse effect on both time and cost. In addition a contractor who seeks to make profit will not want to incur the cost of re-working or re-building to correct faults, and will work efficiently and effectively to produce good quality by 'getting things right the first time'. A profit-conscious contractor will also wish to complete on or ahead of target dates in order to get on with the next contract. This line of thinking does illustrate an aspect of project management that will be discussed more fully in Chapter 10, namely the concept of team-work. A building contract should not be approached as a contest between participants whose primary objectives are in conflict. Each party to a contract will have distinctive objectives which may have to be the subject of agreed compromise in some details but need not conflict. It is this project management approach which has been behind the thinking of some of the new forms of contract introduced in recent years; some of these are discussed in Chapter 11.

Project beginning and end

In the case of every project it is possible to establish a time before any part of it had commenced, i.e. before the 'concept phase'. Similarly there is a time after which nothing has to be done on the project, i.e. after the 'completion phase'. In such a case the 'project life cycle' will have been completed. While this seems to be a simple statement it does require further amplification. In the instance of a new building the concept phase will be the recognition by the client organisation that it has a need for a new building to fulfil a specific purpose. This might for example be the need for a building to house a range of new equipment which will be required to produce a proposed new product. The end of this project from the client's viewpoint would be when the new production facility is in position with its new equipment installed and operating, and possibly also stocks of new product lying in customers' warehouses. It is clear that this manufacturing company is interested in the whole of the project life cycle which comprises all the steps between the initial concept and the commencement of normal operation of the plant. It may however only have primary responsibility for the preparation of the brief, appointment of the design team, the purchase of the plant and the setting up of the operating organisation — no mean task in itself. The architect will probably have responsibility for the co-ordination of the brief, the building design, possibly the award of the contract and its supervision, and the hand-over to the client. The contractor will be responsible for the construction work and possibly for co-ordinating other contractors involved in the plant installation. It can therefore be seen that here we have projects within projects within projects, rather like a nest of Russian dolls. Each of these projects will have to be clearly defined and managed, and each will have successively short time durations. One of the implied overlying problems here is that the client is primarily interested in getting from beginning to end as quickly as possible, and may become a little impatient about the long time lapse between the two, especially as there may be limited scope for the client to influence progress during the intervening stages. An important aspect of the interconnection between a series of projects in this way is the need for careful definition of the scope of each project and a clear statement as to where the project management responsibility for it lies. Definition of scope and responsibility are covered in some detail in Chapter 3.

Projects are made up of a large number of tasks

This is a statement which at first would seem to be true of production of any reasonably complex product. The essential feature in a project is that most of these tasks will be non-repetitive and each will be dissimilar from other tasks which precede or follow it. In fact it is often the case that adjacent tasks are carried out by different groups of people. Chapter 2 discusses the differences between various types of production, but it is relevant here to illustrate the above point with reference to the construction industry. The building of a school will consist of many separate tasks, starting with excavation for drainage and foundations, then concrete footings for walls, followed by brickwork to ground floor level and so on. Each of these tasks will be performed by a different skill squad, and will not be repeated. It is very important however to note that there is much interdependence between these tasks; they must be carried out in the correct sequence, and the scope for overlapping them is very limited. There is at the same time a need to avoid an undue delay between the tasks, so that the project progress is not unnecessarily restricted. By comparison the production of the constituent materials of construction is carried out in a very different way. Cement for the foundations is made in a vast production plant which produces the same product day in and day out; the same people do the same job every day and the product is uniform and consistent. The cement made can be stored until it is wanted on site. The same is largely true of the production of bricks, but the manufacture of windows is more complex. In this case there will be a number of different tasks to carry out, but any one operative or machine will undertake a repetitive task. Final assembly may be rather more like a little project, but even then the whole assembly will be repeated many times. The most important difference is that in a project the particular combination and sequence of tasks is unique, i.e. it has never been done exactly before and will not be repeated again.

Projects are unique

There may be a few exceptions to this statement, but it is almost always true. Every school or hospital will be different, and even in the case of houses it is likely that there will be a number of differences in relation to the particular site, although it must be said that in the case of house-building there are often many sequences of work which are repeated and hence lend themselves to the methods of management used in factory production; this is very much a special case however. The uniqueness of nearly all projects has been one of the main reasons for the evolution of project management methods, simply because there is no exact precedent for any project which will have served as the basis of 'learning by experience'. In factory production it is possible to make one unit of product and observe very carefully both the methods used and the end result. This provides a great deal of experience and information which can be used to improve performance next time round. Methods can be modified or changed totally, and several iterative cycles performed to perfect the operation; any defective products so made can be rejected at little cost, since this can be recovered from the many thousands or even millions of successful cycles which will follow. The phrase 'get it right first time' has been adopted by specialists in quality management and it is easy to understand why. It is however absolutely essential in project management since the unique nature of projects means that there only is one time for each project and experimenting is not feasible. There are many other ways in which the unique nature of projects has had an impact on their management, and these are discussed in Chapters 4 to 7.

Project demand on resources

Given that projects comprise a complex sequence of a large number of interdependent tasks or jobs it is clear that there will be a very complicated demand for a range of resources. The equipment and skills used in the early stages of a project will differ from those needed later, and the demand for any one resource is likely to vary enormously from day to day. This presents real problems for the planning of project work, and is one of the main subjects of Chapter 9. The influence of this variability of resource needs goes beyond the management of any one project; the heavy dependence of the construction industry on hired plant is very much due to this aspect of its work. The design of many projects is influenced by resource needs. The essential difference between projects and steady-state production in relation to resource use is that in projects resources have to be made available when they are needed, whereas in production resources are generally available and the main management task is to keep them fully utilised. There is considerable variation between projects in the extent to which they are influenced or even dominated by the availability of resources, and this has a big impact on how they are managed. This topic is included in the discussion in Chapter 9 at the end of the section on resources.

PEOPLE IN PROJECTS

From the above discussion of characteristics of projects it can be seen that people have a very important role to play. The very fact of uniqueness means that it is not easy to pre-design a planning and control system that will serve to manage any project. Much of the skill in managing projects is in foreseeing and dealing with things which some people might regard as 'unexpected'. The great need for good communication and leadership in a complex situation again puts heavy reliance on the individual project managers. The early work on project management concentrated on methods and techniques for planning and controlling projects, but recent developments have recognised the central role of the project manager, and have concentrated on the attributes of managers and project teams. In his book *Managing Projects in Organisations* (1988) J.D. Frame entitles his third chapter 'Capable people: the heart of every project'. More and more organisations are recognising the truth of this phrase, and this has been the moving force behind the steady expansion in the number of appointments of project managers, and the emergence of relevant education and training programmes. If nothing else project management is all about making new things happen, and this clearly requires conscious decisions and action by people. Placing Chapter 10, 'People in project management' relatively late in this book does not imply a lack of importance given to it, but simply reflects the need to understand something about the subject before considering who should be responsible for carrying it out.

THE DIFFERENT TYPES OF PROJECT

In the academic literature there have been several attempts to classify projects into different types. This is not essential, but it does help readers to put their own project into the appropriate context, since in some ways this may affect the approach used. Briner et al. (1990) discuss this in relation to the people in project management and make reference to some, but not all, of the following groups, and use slightly different meanings to some of the terms.

7

'Hardware' projects

These are where the project comprises the construction of some physical object, such as a building, an oil refinery or a ship. Such examples are all large in size, but other smaller hardware projects would be the stripping down and re-assembly of an aircraft engine etc. The main characteristic is that physical operations have to be carried out, and will usually have high costs of labour, plant and materials which greatly exceed the actual management costs.

'Paper' or 'soft' projects

These projects are those where no physical construction takes place, but there are many interacting tasks, e.g. the creation of a new organisation in an overseas country to undertake the selling and distribution of a company's products. Similarly the design of a new building or of a computer system can be regarded as a paper project, even though each of these may be coordinated respectively with the actual construction of the building or installation of equipment, each of which consists of hardware. In the new age of information the number of soft projects is steadily expanding.

Note that Briner et al. (1990) include both hardware and paper projects in a single group they call 'concrete', with the most important feature being that such projects are under the full-time management of a project team which has no other responsibility. This may suit the consideration of team-work in projects, but the separation between hardware and paper has significant relevance in other areas, such as resources and finance. The term 'concrete' is therefore not used in this book as it could be mistakenly regarded as synonymous with 'hardware'.

'Temporary' or 'short-term' projects

These projects are often of 'soft' nature, but are really distinguished by the fact that they are usually under the control of a part-time project team set up from within an organisation for this project only. Individual members will probably still spend the majority of their time on their regular jobs in design, manufacturing, accounts etc. Typical projects in this group will the acquisition, installation and commissioning of a new production machine, a new control system, or a new ordering procedure. In a sense many major projects are only temporary, e.g. a power station has a finite project life covering design and construction, and is therefore temporary rather than permanent. It is the project management team in the case of temporary projects which is really temporary or short-term and which has implications for the techniques used. Hardware and other large projects, even though they have a finite life, will probably be managed by a team of career project managers who have few other responsibilities.

'Open' projects

Although these are referred to by Briner et al. (1990) they do not really lend themselves to the methods of project management. They are actually on-going programmes with no specified ending and frequently no clearly defined objective. They may consist of a continuous search for new markets, possible new product areas, improvements in productivity, or quality or safety. These are all highly commendable activities, but their approach is somewhat different from that of project management as it is now widely understood. Clearly the search for improved productivity or quality must always be given serious effort, but it would be wrong to subject it to specific time or cost limits. Any particular opportunities for improvement which may arise from a search programme may in themselves become hard or soft projects in the meanings stated above, and thereafter be planned and controlled using the techniques of project management.

PROJECT MANAGEMENT — 'THE BODY OF KNOWLEDGE'

Many people have sought to define the subject or discipline of project management, but it is very difficult to give a definition which is succinct, comprehensive and comprehensible. It is one of the purposes of this chapter and indeed of the book as a whole to put together the essential ideas and principles of the subject in such a way that a comprehension is gained in a gradual and overall way. It is not thought to be feasible to convey the approach in a single sentence. This difficulty of definition has led to the preparation by the UK Association for Project Management (APM, formerly known as the Association of Project Managers) of a document of some 60 pages which sets out what it sees as '... the essence of ... the knowledge and experience that people involved in the formal management of projects would have or need to have'. While this statement does not explain what project management is about it does perhaps serve to define the context of the subject, and does have the advantage of carrying the authority of one of the world's leading professional bodies in the field of project management. One of the reasons why the APM prepared the *Body of Knowledge* (1996) was to guide the education and training of project managers, one of the main objectives of professional associations. It was the absence of a generally accepted definition that made it necessary to have a document which sets out the full range of topics to be covered in a training programme together with a description of the working experience of those who seek professional recognition. It should be emphasised here that project management is not only for those who wish to become professionally qualified as project managers, but also for the many others who work in a situation where project management is being used. A comparison here is that it is not only bank managers who are involved in the management of banks, not only football managers involved in the management of football etc. Many specialists will be engaged in building projects; architects, structural mechanical and electrical engineers, quantity surveyors, builders, accountants, health and safety experts and so on — all will be part of the project management team and will need to understand its general approach. This book will keep on using the phrase 'project management approach' with its emphasis on one-off production and all that this implies for management. All of the people referred to in the above sentence must have this general understanding of projects if they are to make a positive contribution to the project in which they are engaged. It is therefore of relevance and importance that all who work in projects should have a good basic grounding of the subject, possibly at one of the lower levels set out in APM's *Body of Knowledge*.

TYPICAL EXAMPLES OF PROJECTS

The following series of thumbnail sketches is intended simply to illustrate the way a project can be recognised in terms of the application of the approaches and methods of project management.

New superstore for one of the national groups. In addition to the actual design and construction of the building for such a project, there will be many other tasks to be performed. These will include market surveys, financial appraisal, planning and other consents, local supply and service agreements, staff recruitment and training, transport arrangements, insurances, and many more. Each of these will be essential components of the project, each will be somewhat different from what was done in any previous project, and each will take time and consume resources. All of these tasks must be performed correctly or the store will not function properly or possibly not at all. It therefore makes sense that the tasks should all be planned and controlled in every aspect — that is project management!

Introduction of a comprehensive computer system to an owner-managed medium-sized shop. This will not simply mean the selection of a computer, its purchase together with some software, its installation and a couple of days training for the cashier. First decisions must be taken on what the system will be expected to do, e.g. does it include stock control/re-order procedures; this will in turn involve some study of needs and capabilities. The selection of a suitable consultant and/or supplier will be important and will be time-consuming. Once a set-up has been designed there will be all the peripheral activities such as re-order systems, special stationery, accounting procedures, selection and training of staff; again all these tasks and others will take effort and time, probably in this case by people not familiar with either projects or computer systems, and still with their everyday management jobs to be done. Even though the total value of a project such as this will not be very large, it will be a fairly big investment for a small business, especially in the efforts of its manager. If it is allowed to drift on with a sort of crisis management approach, tackling each task only when it becomes urgent, this will be a recipe for trouble if not disaster. It is unfortunate that many small and medium-sized businesses see the prospect of change as a possible crisis rather than a real opportunity for advancement; in fact many will only face up to change when it is forced upon them, and in some cases that may be too late. Change does not necessarily mean crisis, and a way of looking at this is to say that 'one person's crisis is another's project', but only when it is properly managed. Many computer installations have run into trouble not for reasons of inherent system failure, but because of the lack of proper overall management of the project. If all aspects are properly considered, with proper definition of an objective, and careful planning, execution and control of all the tasks, then there are good prospects that all will be well. There are always possibilities of the unforeseen happening, but even then good project management will examine the risks inherent in the project and will take precautions to avoid them or to minimise the consequences. Many such projects have been the subject of books (Donaldson, 1978; Gildersleeve, 1985; Lewin, 1988; Price, 1986; Rakos, 1990).

Introduction of a new product. For many organisations the introduction of a new product will be something that is undertaken only once in a while. This means that it is not the full-time activity of a special group of people, and hence can be thought of as a 'temporary' project in the terms discussed previously on page 8. The setting up of project teams is discussed more fully in Chapter 10, but clearly it makes sense to give the responsibility for the management of a project like this to a special team. The team will have to make an evaluation of the proposed product or may even have to conduct the search for a new product if the company's management has simply recognised the need for new areas of business, and has decided on diversification for strategic reasons. Many large companies have continuing programmes for new products, and in these cases it is often a matter of deciding among a number of alternatives competing for resources. Such organisations have developed a range of techniques in areas like investment appraisal, project selection, and a rather special form of project management which deals with development projects where the outcome cannot be known in advance. Such 'uncertain' projects require a somewhat different approach, and this is discussed as a special topic in Chapter 13. For the smaller company however the introduction of a new product lends itself well to the standard techniques of project management. In terms of project type discussed earlier this may be regarded as a 'soft' project up to the stage where physical work is started on new production facilities. All the earlier stages of market research, product design, evaluation, prototypes and so on will not have used production resources but will certainly have taken time and will have required management. It is best if this is properly seen in terms of project management.

Major plant overhaul. The annual maintenance shut-down of a large plant installation such as an oil refinery or petrochemical plant presents a special case of project management, especially as the minimisation of production loss is very important. This is due to the fact that such 'heavy' industry is capital-intensive, and it is vital that the plant is in operation as much as possible; it is common for such plant to be classed as 'continuous operation' which is literally true in that it is expected to run night and day, seven days a week for periods of eighteen months or more. Much thought is therefore put into the planning of the shut-down project, with the emphasis on minimisation of project duration. Tasks are broken down in great detail, some of them requiring perhaps only a few minutes' work. Alternative procedures are examined before the project is initiated, and regular updates are made as the work progresses, possibly several times a day. In the desire to avoid delays due to the shortage of resources it is usual to ensure that all equipment and skills which may be called upon during the project are available on a stand-by basis; this may be uneconomic in terms of resource use, but it avoids expensive delays. One of the major problems in this type of project is the uncertainty of what might be found when the plant is opened up for inspection, e.g. unexpected excessive wear in some of the moving parts, more corrosion than anticipated, and so on. The project team has to be prepared to meet all of these events, and still get the plant back into operation as soon as is possible. There is a contrast here with the building industry where all information should be available when it is needed, with the possible exception of work below ground, which cannot always be fully foreseen. It does of course remain one of the challenges to the building industry to ensure that information is made available on time, and here the integration of design and construction under the control of one project management team can provide the answer; this does not necessarily mean a design-and-build contract, although this is one approach. The coordination of the contributions of several separate independent parties to a project can be properly managed, but the various responsibilities and authorities must be fully defined. This is the subject of Chapter 9 in which the organisation of projects and the formal documentation of them is discussed.

Development of innovative new plant. From time to time organisations are faced with a major problem which seems to defy a simple solution and requires significant research and development. It has already been stated that many research and development (R&D) projects require a special approach to their management, because some programmes are simply set up to seek new products to expand the company's product range, or to replace obsolete products, and therefore have a general rather than one specific objective. As already stated these are regarded as being covered by programme management, which is a special extension of project management. The case presented below is typical of the problem-oriented project, where a need was recognised by the chief executive of a major construction group.

For many years the standard method of driving steel, concrete, or timber piles into the ground was simply to hit them with a hammer, albeit a large power-driven one. This inevitably causes a lot of noise, which is of considerable nuisance to everyone, but also sets up ground vibrations which can cause physical damage. The actual motivation to set up a development project was partly the altruistic one of noise reduction, but partly to avoid restrictions placed on the hours of work, and the potential for legal actions for damage caused. The stated objective for the project was to develop a 'silent' pile-driver within a period of two years. The means by which the problem was tackled does not perhaps lie within the field of project management, but it is worth diverting for a few moments. Most of the past attempts to quieten construction processes have been based on surrounding the equipment with some form of muffler, but this can only be partly effective. Other more recent approaches have included the introduction of 'negative sound' where vibrations are deliberately introduced that exactly oppose the vibrations of the equipment and hence cancel it out. In the case of the pile-driver however the concept of lateral thinking was applied; it was the hammer which caused the noise and it should therefore be eliminated. An alternative to hitting the pile was to push it, and thereafter the project simply became a matter of devising the method and perfecting the design and operation of a pushing machine; that however is a case for equipment design rather than project management. As a postscript to the case it is interesting to note that a competitor was also attacking the same problem by using a high-frequency vibration system, and it so happened that their system was best suited to gravel conditions while the 'pusher' worked best in clay. The final comment on the case is that it represented a specific problem with a single clear objective, and had the backing of adequate resources to find a solution; it therefore lent itself to the techniques of project management.

Many other projects could be cited, e.g. in high-technology projects (Badiru, 1988); in computer systems (Bentley, 1982; Donaldson, 1978); and in research and development (Bergen, 1986).

On-going projects. There are some so-called projects which have a continuing existence with no clearly identified objective, but rather a general statement of an aim; for example in a production plant there may be a 'project' seeking the continuous improvement of quality, but this is more truly a 'programme'. In fact it is likely that such an undertaking is not referred to as a project, but the group responsible for it may be called a 'project team'. This is an inconsistency of terminology which is unfortunate especially as project management insists on clarity of definition and communication.

2   What is management?

It is not the purpose of this chapter to try to discuss the whole range of management theory and practice, but rather to look briefly at those aspects of management which are relevant to produce and/or sell goods or services to clients. To be more precise it seeks to discuss the methods of management which are relevant to production, and to examine how this differs between project management and the other major area which is usually referred to as production or operations management. It should be emphasised that these two approaches to management are not mutually exclusive, and there will be many situations in projects where the techniques of production management can be useful, and conversely many cases where project management methods can be applied in operations management. Nonetheless there are differences of approach, emphasis and context, and these are worthy of exploration. Chapter 1 gave an answer to the question 'What is a project?' and an introduction to a number of examples. In order to make the comparison with steady-state production it is now necessary to state what is meant here by that phrase. The essential factor is that in production a product (or sometimes a service) is made for sale and/or distribution, with the use of resources which are essentially fixed, and with the efforts of an established and continuing labour force. It is normally the case that the product will be repeated many times, and it is usual for it to be made in anticipation of future sales rather than in response to a specific instruction. In order to further explain the differences between production management and project management it is helpful to classify products in two separate and simple ways, rather than by using the standard industrial classification which divides the whole of industry up purely on the basis of the product made, e.g. chemicals, construction, computers, pharmaceuticals, agriculture, and so on. The two areas of classification are (i) product type and (ii) method of production.

PRODUCT TYPES

A very simple division of all products can be made into two groups, namely 'integral' products and 'dimensional' products. An integral product is one which comes in whole units, often in fact in an 'integrated' assembly, made up of a number of component parts. It is usual for such products to be counted in whole numbers (i.e.

14

integers), for example thousands of motor cars from a car assembly plant. In contrast the so-called 'dimensional' products are those which are homogeneous, often in the form of a fluid or powder, e.g. oil, cement, paint; or perhaps in a continuous length, such as paper, wire, cloth. The characteristic feature of dimensional products is that they are measured not by counting but by observing an actual dimension; this may be weight (e.g. cement), volume (e.g. oil) or length (e.g. cloth). This division may seem to be arbitrary, but it does have significance for some of the management controls which are relevant. Consider first the question of fulfilling an order; it is likely that it is not essential to have extremely accurate measurement of the quantity of a substance being bought. For example, if an order for 10 t of steel is placed but for some reason the delivery is short by 1 kg, it is unlikely that it will be of much consequence, and in fact it is most improbable that anyone would ever know. By comparison if an order is placed for a lorry weighing 10 t and it is delivered minus a component which weighs 1 kg, there is a real possibility that the consequences could be very serious, and there is a high probability that the absence of the part would be noticed fairly quickly. While this is all very obvious, it does point out the need for a different approach to the checking of out-going goods. In a similar way the matter of quality should be thought about. In a dimensional product its homogeneity means that there is likely to be consistency of the parameters of quality, or at worst a gradual change with time. In the particular case of liquids and other fluids it is possible to improve the effectiveness of sampling, simply by stirring up the whole batch of a product for an adequate period and testing only a very small sample. On the other hand putting a whole batch of electrical components into a drum and stirring up the contents prior to testing a sample of a single unit will tell you relatively little about the characteristics of the whole batch; a much more sophisticated statistical technique is necessary. In a different way the correction of faults may be easier in an integral product especially if a fault is simply in one component which can be replaced; this may be more difficult in a dimensional product since any fault is likely to affect the whole batch; again we are presented with a different problem to tackle.

METHODS OF PRODUCTION

Here three broad areas of classification are used, namely jobbing work, batch production, and mass (or continuous) production.

Jobbing work

Jobbing work is the general method of producing one-off items, usually on the specific instructions of a customer. The actual word 'jobbing' is really only used in a few special cases, e.g. a jobbing builder is usually a small firm which will undertake small works such as house extensions or repairs. A comparable product would be a 'made-to-measure' suit ordered from a small tailor who will personally cut and sew a suit to the customer's size, shape and style requirements; this is usually referred to as bespoke tailoring. The concept is more widely applied and is by no means confined to small firms; a major contractor constructing a power station is also carrying out one-off production on the specific instructions of a client, and for the purpose of our discussion is therefore carrying out jobbing production. It is nonetheless clear that Bovis, Taylor Woodrow, McAlpine and others would not like to be referred to as 'jobbing builders'! Therefore, it might be more appropriate to call this type of work 'project work' which is really what it is. The general approach is not confined to the examples quoted above; there is a great deal of work done by engineering firms to make special items of plant to be incorporated into production equipment. In publishing, each book could be regarded as jobbing, even though the actual printing of it will be by a batch process.

The characteristics of jobbing or project work are various. Each item of production is unique and is made either to the client's own instruction or to a design which has been agreed in advance. This means that products are not made speculatively with the hope and intention of subsequently selling them, but a sale is assured by the time production commences. Consequently, the manufacturer does not have stocks of goods for sale with the consequent problems of cash tied up, ware-housing, obsolescence and the associated risks. The approach to sales is quite different, usually entailing direct dealing with the ultimate client rather than through a wholesaler and/ or retailer. It does present problems in that samples cannot be shown of the exact product and the customer will have to wait for the item to be made. There are many other areas of difference which have implications for management, but perhaps the most important for the present discussion are in the area of actual production.

The production equipment in jobbing or project work is usually of a fairly versatile nature compared with that in other types of production. The tailor will have scissors, sewing machine and a press, the jobbing carpenter or joiner will have an array of hand tools plus power saws, drills, routers; all of these are very versatile tools and mostly simple and hand-controlled. This last aspect is crucial in the management of the production process; the fact that tools are operated by hand implies that the real control or influence over the process in terms of speed, accuracy and effectiveness is literally in the hands of the operative. This means that management control over the processes of production has to be exerted through the operative, usually by a supervisor or foreman, who in turn has to be supervised. The implications of this in terms of the selection and training of the whole workforce and its organisational structure are clear. Unfortunately the training background of the construction industry is not uniformly good, and many of the problems of that industry can be traced back to inadequacies in training. Even in the case of large-scale jobbing or project work, e.g. major construction, it is true that most of the tools are hand-controlled. It is probably an over-simplification but it is sometimes said that there are only three main operations in construction; placing things, cutting things and hitting things. This is easily seen in the case of the carpenter/joiner and the bricklayer, but it is also true in heavy construction where placing is frequently by crane, an excavator is just a large cutting shovel, and a pile-driver is a big hammer! The analogies may seem a little far-fetched, but the important point is that all of these items of plant are hand-controlled by one person operating a series of levers and pedals. The opportunity for error or even just less than perfect execution is great, and presents a challenge to site management. It is not only a question of training and supervision, but, especially where the detail of work changes from day to day or even from hour to hour, there is the problem of conveying exact instructions to the operator who may be some distance away from the supervisor and out of visual contact.

Another aspect of the versatile nature of equipment in jobbing work and the uniqueness of each product is the great variability of demand for that equipment. It becomes very difficult to balance the workload for various items of equipment, and this can lead to conflicting management requirements. In a simple jobbing joinery shop it can be expected that whenever there is a need to cut some lengths of timber there will be a power saw available to do the job. The principle on which such a shop would operate would be to have enough simple machines available to ensure that no-one is kept idle waiting for a machine; all machines are therefore likely to have a fair amount of idle time. While this is acceptable for simple tools for which there is a big demand, it becomes doubtful whether more complex and expensive machines can be regarded in the same way. The manager of a jobbing shop is likely to be concerned if expensive equipment has too much idle time, and this becomes increasingly true on large projects, although there are some cases where possessing very large and expensive machines is justified even when they get little use. A good example of this was the construction of some of the early nuclear power stations which used enormous 'goliath' cranes capable of lifting 450 t to a height of 70 m and transporting them up to 500 m. Such lifts were few in number, only 16 on at least one station, but the cost was justified by the very large saving which was made in the total construction time. It is just not possible in jobbing work to have equipment which is always available whenever wanted, and at the same time is fully utilised. This is one of the facts of life that the project manager has to learn to live with and to manage. Many managers of steady-state production just do not seem to understand this problem.

It is already clear that most project work will fall into the category of integral products made by a jobbing process, but it is worth looking at the other process methods to see the way these influence the management of production.

Batch production

As is clear from its name the practice here is to make a quantity of a product of consistent form and then move on to another similar but not identical product for the next batch. A very simple example of this is in the manufacture of paint, where a batch of one colour is made, and then a batch of a different colour, made using the same equipment and procedures. In the case of paint there is the obvious need to clean the plant out between batches, and in other processes there will be corresponding changeover tasks to perform, such as fitting new cutting tools, altering machine settings, finding different component parts etc. Some of the specific problems to be faced by managers of batch production are different from those in jobbing work, e.g. how big the batch size should be. There are well-known techniques for tackling this question; while there are economies to be made by having large batches which minimise the changeover costs, it does mean that large stocks of finished product will have to be kept, involving space, money and the risk of deterioration and obsolescence. Most products made by batch production will not be made to customers order, but will either be for sale to the public or used as an input to another production process. This involves all the tasks of marketing and selling, anticipating future markets, and making decisions on the purchase of manufacturing plant. Such decisions are important because the plant is usually of a fairly specialised nature which means that it cannot be used for any other product. Also it is often relatively expensive and must have a high utilisation to be economical. In the two cases of integral and dimensional products all of these factors have given rise to fairly sophisticated production planning techniques which have to be used to cope with unexpected demands for different products, and the even more difficult linking together of a number of separate batch processes which have to be used in sequence to make a range of products. These are very different problems from those faced by a project manager who may simply call up whatever plant is needed at the appropriate time, provided of course that the need has been anticipated. The scheduling of batch work is also significantly different from the case of mass production, to which reference is made below.

The question of where control lies was stated in relation to jobbing work to be often literally in the hands of the operative. This is not quite so true in batch work because many more of the actual operations are automated. For example, a machine tool will be preset to make components of a specified size and shape; there is considerable time and skill required to set the machine up at the start of a batch, but once it is done it can be checked and the item produced also checked before a batch is run off. Process temperatures, quantities of raw materials and other decisions may have to be made in batch chemical production. There is still therefore considerable direct input by operatives in batch processes, but it is reduced because of the more repetitive nature of the process, with fewer new decisions needed than is the case in jobbing work. The relative location of process control is indicated in Fig. 1.

Feature of production ▲ Predetermined central control O Local control	Jobbing (project)	Batch	Mass/ continuous
What is produced?	▲	▲	▲
How is it produced?	O	▲	▲
How much is produced?	O	O	▲
How quality is achieved	O	O	▲
How quickly work is done	O	O	▲
How much work costs	O	O	▲ / O

Fig. 1. Where does control lie in the production process?

Mass production and continuous production

In this category of production process these two phrases have distinct applications; 'mass' production usually refers to component manufacture and assembly which relates to integral products in the terminology used here. The term 'continuous' production means exactly what it says, namely that the process operates 24 hours per day, 7 days per week, and for many weeks at a time, perhaps as many as 100. By their nature, continuous processes are usually concerned with the manufacture of dimensional products. One of the most important features of this category of production is that the plant is totally dedicated to the manufacture of one product, with only very limited variation in that product. The plant is nearly always very capital-intensive, which means that it is important that the plant should be kept in operation all the time. In the special case of continuous processes this is compounded by the very difficult and expensive start-up procedures which make it imperative that stoppages are avoided. Hence, there is great pressure on management to keep the plant running at almost any cost; this is partly achieved by continuous monitoring of the plant and the product, and having spare components and squads of fitters standing by 'just in case'. In other types of production such a policy would be excessively expensive, but in this case the cost of having plant idle is even greater. A more detailed discussion of this topic has been given in an earlier book (Woodward, 1982).

Returning to the matter of process control reference is again made to Fig. 1. In mass or continuous processes the plant is not only capital-intensive, but is also very complex. It is designed to make only one product, usually to a very tight specification. Many of the process controls are built into the plant, often with computer on-line control, leaving the operator with little freedom to change anything but simply to ensure that all is well. Even this role is eliminated in automated or robot-controlled plants which run virtually without human intervention. All of this shifts the locus of control away from the plant operation and back to the design and construction of the plant.

Summary of product and process classification

Table 1 summarises the preceding paragraphs and gives some examples of typical products.

Table 1. Typical products of the main methods of production

Production method	Product type	
	Integral products	Dimensional products
Jobbing work (multi-purpose tools)	Project work, e.g. construction	Effectively none
Batch production (specialised tools)	Pumps, motors	Paint, beer
Mass production	Motor vehicles, cooker, TVs	
Continuous production		Cement, petrochemicals

Table 2. Jobbing and mass production — where control lies

Control over	Jobbing	Batch	Mass/continuous
What is made	Contracts manager	Board	Board
How much is made	Contracts manager	Board	Board
How is it made	Project manager	Production manager	Chief executive
Cost	Supervisor	Production manager	Chief executive
Quality	Operative	Supervisor	Chief executive

In some of the paragraphs above, reference has been made to the ways in which product type and production method influence the management of production, in particular noting where control over the process lies; this is summarised in Table 2, which indicates at which level of management the main influence is exerted.

PROJECT WORK WITHIN THE CONSTRUCTION INDUSTRY

Project work has already been stated to lie firmly in the category of integral products using jobbing production methods, and given that nearly all construction work is project-based it follows that most forms of construction also fall into this area. It has also been noted however that most of the components and materials used in construction will have been produced by one of the other production methods, either, batch, mass or continuous. It is useful to look at various aspects of the construction industry to see where it differs from other industries, and especially where these differences are of significance for management of the industry.

Uniqueness of the product

This paragraph is written in the specific context of the construction industry, but it is applicable to many other project situations, especially where there is a design element included. Most building or other construction projects are individually designed, and in fact the design is often regarded as part of the project. This may have the effect that the design is not complete at the time of the start on site, which can have serious consequences if it is not properly managed. It also means that a very significant part of the project duration is taken up on design, and sometimes clients do not understand why they do not see progress on site at an earlier date. This can lead to pressure on both the design team and the construction team to get ahead, and this desire to 'get on with it' can lead to expensive errors being made. There is a definite need to plan ahead. There have been famous examples of massive holes being excavated in the wrong field, simply because the site team had the machines available, good weather in prospect, and the enthusiasm to begin, even though final clearance of the design had not been given. Mistakes such as this are not easy to put right — you cannot easily throw away a large hole in the ground!

It is not only the time taken for the design that presents a potential problem; the fact that the design will only be used on that one project means that the design effort represents a significant part of the total cost, as the cost of design cannot be spread over a large number of units of production as would be the case in mass production. It also means there is a reluctance to revise the design extensively. For example, suppose the designer of a building states that a major revision of the design would offer an economy in construction of £1000 but would cost £2000 to complete (quite apart from any delay). It is most unlikely that the design revision would be authorised. By comparison, if the designer of a household machine offered £10 saving on the cost of making a washing machine by a design and tool change costing £50 000, the change would be justified if the production run was in excess of 5000 units.

Another difficulty of the design taking place at the same time as construction, is that there is a temptation to change the design as work proceeds on site. Often a client will quite reasonably want to take advantage of new developments which arise during a project, more usually in connection with the inclusion of new facilities rather than changes in building technology, although that may sometimes arise. In such cases a client may believe that simply because part of the work has not been started it should be a trivial matter to make a change. For example, the decision to change from one type of press to another could mean that it is necessary to use a steel frame for a building rather than load-bearing brickwork. This would clearly have implications for foundations and would perhaps introduce a new sub-contractor with whom a price would have to be agreed before making the change. The potential for disruption of contract progress is significant. This does not mean that change to design must always be resisted once construction has started; it is always important to remember that the client is paying for the project, and therefore should be able to make changes if these are thought to be desirable. What is important is that the client should be made to understand the consequences of any change, so that a reasoned decision can be based on what the consequences of the change would be, and whether the impact on costs and project duration are justified. It should be perfectly possible to make such a judgement in the case of construction and other similar long-duration projects. It may be more difficult in the high-technology industries of computing and information management, where the rate of change of new ideas is so rapid. In the management of design development it is always notoriously difficult to know when to 'freeze' a design and go into production with a product; this happens in batch or mass production when much investment has to be committed to capital equipment, but there always remains the option to change the product at any time in its life, albeit at a cost. In project work however the design is only completed once, and the product only made once; it will therefore have to survive its whole working life based on that original design. Hence the pressure during a project to take advantage of technology as it develops. Yet another aspect of the uniqueness of the design of a project is that by definition as a one-off it cannot have had all its 'bugs' identified and eliminated. A unique design for a building will not only be untried in its totality, but quite often the form of construction may be new and untried. Where this arises, time allowances should be built into the programme to allow for unexpected difficulties, in much the same way that contingency periods should be incorporated for delays caused by bad weather etc. In various ways it can be seen that the unique design associated with projects does have a big influence on the way that design is managed.

Choice of method of production

A fundamental difference between jobbing and the other forms of production is that the actual method of production is decided largely by the people doing the work. In mass or continuous production the design of both the product and the plant producing it are closely integrated. Batch production plant is to some extent more versatile, but it is only in project work that the people actually carrying out the work will make decisions on how it is to be done, and what equipment will be used. This can clearly be seen in the case of building where the site manager will make day to day decisions on where each squad of workers is to be employed, what items of plant will be used for each task, the extent to which work will be supervised, what safety arrangements have to be made, the assessment of bonus targets for production and so on. All of these decisions will have an influence on the quality of work done, its speed and its cost. While many of these decisions will be based on the general experience of the staff involved, there may be a number of instances where a new method has to be tried, and no prior experience exists. This carries with it a number of risks, not only ones of a 'go, no-go' nature where total failure of a method may result, but more often where a new method may turn out to be slower, more expensive or less effective than had been hoped. This can inhibit trying out new approaches, because in a system where the product is only produced once there is no benefit to be obtained from a method that needs to be refined several times before it works properly. Sometimes in project work new methods have to be found because the particular circumstances of the project make it necessary. This is frequently true of construction where there is often a desire to build ever bigger. Many of the really big construction projects do present such large problems that the design and construction methods do have to be fully integrated, much as in the process industries, e.g. petrochemicals; these are relatively rare occasions however and are confined to large innovative projects. In any project where production methods are not fully predictable there is the possibility that quality, speed and cost are unpredictable, and allowance must be made for this in planning and controlling projects.

Quality management

Again this paragraph is written in the context of construction but has many parallels in other types of project work. In any one-off project the required quality standards have to be established by preparing a fully detailed specification, rather than by offering samples to be copied. Samples are sometimes used in construction for standards of bricklaying, jointing etc. but it is somewhat meaningless to try to show by sample the trueness in verticality of a wall. Similarly it would be meaningless in a computer project to try to convey reliability by any means other than specification.

One of the problems is that specifications are susceptible to variation of interpretation by different people, and have to be very carefully written. In terms of quality control the popular techniques such as sampling, warning and safety limits, trends and so on have limited application in projects which are only completed once. There is some opportunity to use sampling procedures in the materials and components built into a project, but not in the main project activities. There is one additional area of difficulty in construction and that is what to do with defective work that does arise. It is usually difficult if not impossible to throw away defective items, the very least that has to be done is demolition and reconstruction, both of which incur cost and time penalties. Re-work is sometimes an option but is rarely totally satisfactory, and often has to be grudgingly accepted. The cost of removal and re-building may exceed the original cost many times over, and is therefore often resisted by site personnel. For example, if a hole is excavated in the wrong place it may have to be refilled with concrete rather than with excavated material because of long-term settlement problems. Even this may not be acceptable if subsequent operations have to be carried out in the same area of ground. The subject area of quality is extremely important and requires a special approach in project work; Chapter 7 is devoted to the way in which quality management can be tackled, with sections referring specifically to construction.

Variability of environment

Two types of environment are discussed here; the physical environment in which the project is actually located, and the social, legal and economic environment in which it is being undertaken. Many projects, especially if they are of the 'soft' or 'paper' type, will not be influenced greatly by the physical environment, but others, such as construction will be very much affected by it. Within areas such as construction, problems such as hostile local ground and air conditions can generally be allowed for by taking special measures, e.g. temporary enclosure for weather protection. Perhaps more difficult from a management point of view is the probability that conditions will vary, and in a largely unpredictable way. It is very common for claims for extra time to be submitted based on adverse weather, but these are often totally or partly refuted on the basis that a certain amount of bad weather should be anticipated and allowed for in the planning and pricing. In heavy construction such as marine work there are special problems related to tides which can be partly predicted, and to a lesser extent the impact of high winds can be anticipated.

In most projects it should be possible to foresee the effects of the legal requirements relating to the project unless it is of very long duration. Most new legislation is notified well before it comes into effect, but it is important to be aware of the changes in law that may come into effect within the time-scale of the project. Economic changes are not so easy to foresee; material prices and wage rates may change not only within the general national picture, but also locally. If several local projects start at about the same time and demand the same resources the law of supply and demand may cause a rise in prices. In heavy construction this can be guarded against in various ways, e.g. a contractor when approaching a large construction project such as a power station may buy a quarry near to the site, confident that even if they do not win the contract the material will be wanted for the project, and they will be able to profit from supplying it. There may similarly be shortages of skills in any type of project, e.g. computer programmers, systems engineers, or specialists of many sorts.

All of these considerations of environment present special problems in project management somewhat different from those in steady-state production, which is usually carried out in a more stable, or at least more predictable environment.

Method of selling

The methods of selling projects is very different from those employed in selling most other products. As stated earlier nearly every project is put together at the specific request of a client, even if the client and the producer are both departments within the same organisation. It is seldom a matter of making project products speculatively with the intention of making a later sale of the finished item. There have been exceptions to this, e.g. some shipyards which were running out of orders decided to build cargo ships in the hope of selling them, but this was a desperate attempt to keep production going and was not very successful. The make-to-order selling pattern of most projects means that many problems are avoided, e.g. the holding of finished stocks of goods in warehouses and elsewhere. Holding stocks of goods takes up space, ties up working capital, and risks deterioration and obsolescence. The costs of selling and distributing consumer products represent a very significant part of the total costs of such products, again a very different pattern from that in the project-based industries.

The method of selling in the project-based industries is often a two-stage process. It is common practice for clients to invite a number of potential suppliers to submit competitive bids for the project work, based on a specification of what is wanted. It is often the case that companies who are interested in submitting bids are asked to 'pre-qualify', i.e. to demonstrate their ability to undertake the proposed project, usually by reference to previous similar projects and their resources available, namely their 'track record'. The process of submitting a bid for a contract presents its own problems, and these are discussed more fully in Chapter 13; the difficulties lie in predicting the cost of work which has not been done before because it is unique, and also because conditions of work cannot be fully foreseen. There is also the problem of not knowing what price competitors are likely to submit. Selling is thereby another area in which the project-based industries call for a special approach.

Financing of projects

Small projects will usually be paid for by the client once the work is complete and handed over by whoever has undertaken the work. In small projects the producer has to provide all the funding of the work in progress, but is assured of receiving payment as soon as work is complete. This is a little more secure than for manufacturers of many other goods for which a market is not fully assured. In the case of large projects, e.g. in construction, the usual practice is for interim payments to be made by the client to the producer, approximately following the value of work completed. This means that the producer or contractor has only to provide working capital for one or two months work, and to some extent this will be funded by the suppliers of materials and services. The client really carries the financial burden of funding the work throughout the project period; this means that supplying and contracting companies with fairly small availability of finance can undertake reasonably large projects. This in turn brings about a situation which certainly can lead to the relatively high rate of bankruptcies in construction; with only a low need for working capital it is fairly easy to get into the construction business. Because of this fact profit margins are also low and where there is uncertainty about costs it is very easy to get into a situation of an operating loss. Hence it may be easy to get into construction, but it is also very easy to be put out of it. Few other industries are in the same position as those in the production of project-based items or services.

Time span of projects

Projects can range in duration from being very short to being up to ten years. Where they are short there may be the need to undertake a lot of planning in advance of the actual project start, in order to ensure that it runs smoothly. Where project durations are long other problems arise not seen elsewhere. There may well be a change in the brief within a long project period, and it is almost certain that there will be design changes over that period. There may even be major economical or political changes that render the project almost obsolete before it is complete. For example the very large water supply scheme in Northumberland was designed to supply large quantities of water to the steel industry on Tyneside, but the steel industry had closed down by the time the project was complete. The massive projects in the Clyde area to service and operate Trident missile submarines has been completed after the reduction in East–West tensions. This is not a comment on the wisdom of these two projects, but simply a point to illustrate the difficulties which can arise in projects which have a long duration.

The foregoing comparison of project-based and production-based industry discussed in this chapter has been intended to show that there is a big difference between the management of projects and the management of manufacturing industry. It is a myth to think that if you can manage one type of industry you can manage any other. While there may be many similarities there are also many differences. It is the realisation of this fact that has led to the development of project management as a discipline in its own right, with not only a set of techniques but a whole philosophy or approach. This approach is well described by J.R. Turner in *The Handbook of Project-Based Management*. It is the function of this present book to communicate this concept and present the essential material which is the basis of the subject.

Many writers have addressed the question of showing the differences between production management (often referred to as operations management) and project management. Turner (1993) discusses it fully, and Table 3 reflects much of what he says, but is not a copy and has been extended and modified.

Table 3. The differences between project and production management

Projects	Steady-state production (operations)
One-off (i.e. unique)	Products largely repeated
Predetermined total content	No planned end
Activity levels fluctuate	Steady-state
Change the status quo	Maintain the status quo
Creating change is central	Change evolves slowly
Variable resources	Fixed resources
Rolling production team	Static team
Looking forward	Looking back
Foreseeing uncertainty	Minimising variability
Object attainments	Level of achievement

The basic tasks of management

In any production industry one of the major tasks of a manager is the planning and control of the work being done, and it is possible to represent this function by a simple diagram as in Fig. 2.

Planning

Planning is the first step in this process, namely to set out the work that is to be done, so that operations can begin. After the time specified in the plan the amount of work satisfactorily completed is measured, and the actual work done is compared with the

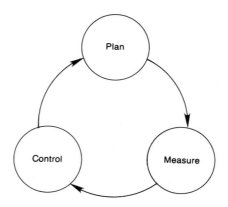

Fig. 2. Model of plan–measure–control cycle

plan. Depending upon this comparison it may be necessary to exert control by taking action. Note carefully the wording here; control is not just the process of measuring performance, which is really 'monitoring', or keeping a check on what has been done. Control means taking action to change something; the only time that action is not taken is when the plan and the performance are exactly the same in every respect and no change is needed. To be in control of a process it is essential to have the authority, power and ability to make the appropriate change to the process. This whole subject will be discussed later in more detail in the section on responsibility and authority in Chapter 11.

Measurement

Measurement is the second step of a manager, and comprises counting or measuring by some means the amount of work completed in the time period. In the case of mass or batch production, measurement is simply a question of counting the number of units produced. In project work it is much more complex, and has to be carefully thought out; this topic is dealt with more fully in Chapters 5, 6 and 7.

Control

Control of some form will have to be exerted by making a change to what is being done. The first thing to determine is the reason for the difference between the plan and what has been achieved, and this may lead immediately to the cause of the difference and the action necessary to correct it. For example, in a production process, whether it is in a factory, office or on a construction site, the output for the week may be 10% down on target, and it is immediately found that this was due to the breakdown of a machine which was fairly quickly repaired. This does not mean that all appropriate action has been taken; an examination of the machine should be made to see if is in need of major overhaul or replacement; possibly provision should be made to have a spare standby machine available to take over in the event of future breakdowns. Another possible action would be to sub-contract that part of the process to another company which has more reliable plant; there may be many options. If however it is found that there is no acceptable remedy for the short-fall in production then it will be necessary to revise the target downwards, and that is the reason for the closing arrow in the diagram from 'control' back to 'plan'. This illustrates the cyclical nature of the control process in production and is often referred to as the 'plan–measure–control cycle'. The same procedure would be followed if the failure to meet target was due to a power failure, shortage of components, an industrial dispute, lack of operatives' effort and so on. The range of possible remedies would be different in each case, but the principle would be the same. There is always the possibility that actual production will exceed the planned target, and again a similar pattern will follow; once you are satisfied that the reason is a valid one which can be relied upon in the future then it would be appropriate to raise the target, if there is a need for increased production, or take advantage of the excess capacity to make savings somewhere.

It is not only in production output that this plan–measure–control cycle can operate. The sales of a product can be viewed in exactly the same way. Sales targets are set and actual sales compared with them. If the figures are below target a number of options are open to management. An increase in advertising, a one-off promotion event, incentives to the sales force, special discounts to retailers, a reduction in price — these are all familiar actions that we see in household goods, but can apply anywhere. Again control is being exerted by taking action to make a change. Similar procedures can be followed in the control of quality, remembering that by quality control we mean having the power and the will to take corrective action. Sometimes the title 'quality control' is used for the process of testing the quality of output from a process, with the limited objective of rejecting defective items; this is not real quality control which should aim to maintain the process at a standard such that defective work is eliminated. This whole subject is fully discussed in Chapter 7, but it is worth stating here that the concept of a 'zero defects' policy is a valid one which is of special importance in the project industry, where there is no opportunity to get it right the second and subsequent times an operation is performed, because it is probably only being performed once. There can be many reasons for a failure to meet specified standards of quality and all possibilities should be looked at when seeking a cure for a failure. There may be occasions when re-examination of the target set may be genuinely the only option to overcome the failure to meet quality standards; sometimes in writing specifications for projects it is easy to simply state the standard that was used in the last project without asking whether it is really appropriate. An example in construction was the requirement for all reinforcement bars in a large concrete structure to be positioned within a tolerance of plus or minus 2 mm. For this to be followed would require a whole new way of fixing and checking reinforcement, at great cost; on examination it was found to have been copied from a specification for highly specialised small precast units.

PROJECT MANAGEMENT AS A DISCIPLINE

The foregoing discussion about the differences between project management and production management illustrates the reasons why a new discipline has evolved to deal with the management of projects. It also perhaps explains why many projects in the past ran over time and over budget. Long ago many major projects were completed on time but this was when the power of the builder over resources was very great, e.g. in times of slave labour and plentiful money and materials. In more recent times the complexity of the commercial and financial environments made the management of projects much more complicated, but the appropriate techniques of planning and control had not been developed. The use of general management methods did prove in many cases to be inadequate, with the result that time and cost targets were allowed to slip. It is still the case that some projects run late and over cost; this is not generally due to the failure of project management as such, but can usually be shown to be due to a failure to use the methods correctly. It is true that many projects are now completed on time and within budget, and that this is frequently, but not always, brought about with the help of good project management. Much of this book therefore is devoted to explaining the essential features of the subject of project management and how it should be used.

THE CONCEPT OF THE TRANSFERABLE MANAGER

It is often found that managers move from one industry to another with relative ease and success. This is seen as a way of bringing in new ideas and approaches, and is often used as a device to enliven an organisation which has become a little stale. However to make a transfer across the project/production divide is more difficult, because of the totally different culture and environment in which the work takes place. It is easy to see that a manager could move from say a food processing company to a clothing company because much of the operating environment will be similar, but for such a manager to take up a post in a construction company would be much more difficult. Similarly, a project manager could perhaps move from a computer installation firm into construction because the culture of one-off work is present in both. It is becoming more widely recognised that project management has a lot to offer in many organisations where significant change is taking place. This is often referred to as 'management by projects', a phrase which is becoming widely used. Most of the techniques which are described in the following chapters can be applied both to the management of projects (as in construction), and the management by projects (as in the case of a company launching a new product). This leads to an apparent contradiction of what was said at the start of this paragraph about managers not being able to cross the project/production divide. What is really being said is that managers can move from one industry to another where their particular function calls either for a project approach or for a steady production approach in both jobs; it is the change from one management culture to another which is difficult, especially if the individual manager does not appreciate the difference, and tries to use the methods of production management in a project environment or vice versa. There are of course many tasks in all forms of management which have similar demands on managers, e.g. the ability to communicate well both in writing and orally, and being a good listener. Problem-solving skills, clarity of thought, initiative and many other attributes are also of general importance, but there are some characteristics which are not the same in projects and production. This and other aspects of people in projects are more fully discussed in Chapter 10.

3 What has to be managed in a project

MAIN AREAS OF ACTIVITY

Having spent considerable time in discussing what a project is, and the way in which its management differs from that of a steady-state enterprise, it is now time to get down to the subject proper. Furthermore, it is possible that some readers will have quickly skipped over the first two chapters in order to get to the meat of the subject. The purpose of this chapter is to set out the main areas of activity which have to be managed in a project. This is done with particular reference to construction, but throughout the chapter separate comments are made to show how what is being said would apply to other situations. There are several distinct areas which have to be managed, and these are listed below. The following list is similar to that quoted by other authors (Turner, 1993) but has been extended and given different emphasis.

- *Scope* What is the project? How is it defined?
- *Procurement* What contractual and organisational route will be used?
- *Planning and progress* Analysing the project, setting a plan of action, and then controlling progress.
- *Time* What is the timetable, and how will it be enforced?
- *Cost* What is the budget, and how will it be controlled?
- *Quality* Determination of standards and their observance.
- *People* Individual project managers and project teams.
- *Risk* What are the risks, who carries them, and how can they be avoided or minimised?
- *Project success/failure*
- *Facilities* The use of capital assets (e.g. buildings) and their maintenance.

It was stated above that these areas can be identified as being distinct, in that it is important that attention is paid to these primary aspects. This does not mean that they are independent of each other; they clearly interact in a number of ways. For example, it is well known that time and cost are each influenced by the other; if it is required to accelerate some work one of the simplest (but often not cheapest) means of doing this is simply to work overtime. On the other hand if something arises to cause a delay in a project there may often be costs involved because workers are left idle until the delay is overcome. This interaction of time and cost is an important aspect of project management and it is fully discussed in Chapter 8, along with the link of quality with time and cost. Let us first look at each of the above topics in more detail.

The scope of a project

It may at first seem obvious that this can be simply stated as being 'to build a school', 'to launch a new product', 'to run an international conference' etc. This is, however, more like a concise mission statement since it does very little to define what the project really includes. In the case of a building project it would be necessary to determine the following and assign responsibilities.

(a) Who is the client? Is this the real client or simply an agent acting for the ultimate owner to whom the building will be transferred on completion? It has been common practice for all buildings to be used by government departments to be commissioned through an agency. Is the named client to be the ultimate user? Many buildings are constructed for development companies who then lease them to tenants on a long-term basis.

(b) Is it clear where funding for the project originates? Even where an agent or developer is identified as the original client they may be obtaining finance from a bank or investment company who will be interested in monitoring where their money is being spent.

(c) Is the project part of a larger strategy or programme of projects? If so how do they tie up if at all?

(d) Is the client an 'intelligent' client? That is, are they in the habit of sponsoring projects regularly, or is this an isolated event?

(e) Are the client and the contractor bound together by some form of partnering agreement?

Answers to these questions will determine how the scope is to be defined and how it relates to other work.

Most building projects will be one-off and purpose-made, and will therefore need to be fully briefed and designed. It is necessary to determine whether these are to be included in the scope of the project under consideration, or whether they form part of a separate project. One recent practice in use in some types of work has been to separate the briefing and design from the actual construction; this has perhaps been done to avoid some of the legal and liability arguments by separating responsibility for the two aspects and making it clear that design is the responsibility of the architect and engineer, while construction is the responsibility of the contractor. This is understandable but it has been found that it does not solve the problems of project delays, because it leaves grey areas which are not properly covered by the two parties, leaving the client to cover them. However, this does lead to an interesting discussion of responsibility in project management which is left to Chapter 11.

The definition of the scope of the project is clearly very important. In the case of a building it will have to spell out carefully the physical site boundaries, possible connections with other buildings, the inclusion of all the appropriate services, the necessary financial and legal requirements and so on. Further coverage of these is given in Chapter 4, here we are simply concerned with the type of activity involved and making the important point that the scope definition is an essential pre-requisite of all the other stages of project implementation. The scope definition is largely a clear statement of what has to be done.

It is possible that the scope of the project will include not only the design and construction of the building, but also its furnishing and even its initial operation. In the case of a factory building it is now common practice for the project to include the purchase and installation of all the production equipment. This may be extended to include the initial purchase of raw materials, the recruitment and training of staff and labour, and even the setting up of transport and distribution systems for the product, together with all the computer hardware and software to manage the production unit. While this may not be a very common practice in the industrialised world it is often required for the establishment of a production plant in a developing country. Even in the case of an 'intelligent' client in a highly industrialised environment there will be a need for the project management of a new production facility; it may not be part of the main contract and not within the scope of the building project, but all the steps listed above will have to be managed, possibly by the client organisation itself in the form of a sub-project within a project.

The development of a new product

The preceding paragraph prompts the concept of a project which does not involve building as such, but does include many of the features which make project management appropriate. Consider the case of an established manufacturing company which has decided to replace one of its existing products by a new one. Let us assume here that no building work is involved, and that all the actual work is to be carried out by the company itself. First the scope of what has to be included in the project must be defined, possibly by asking if the following are within the project or not.

- Market research related to the new product.
- Design of the new product.
- Production methods to be used, i.e. the engineering of the product. (Does this imply decisions on whether the new design will be constrained by existing production plant, and to what extent re-equipping will take place?)
- Sources of materials and components.
- Checking of new and impending legislation relative to the product.
- New standards within the industry.
- Checking patents and licences.
- Packaging and supporting operating instructions.
- Distribution methods.
- Changes in working practices within the plant.
- Operative training, and perhaps recruitment or redundancy.
- After-sales servicing.
- Product launch and promotion.

Some or all of these activities will be included and will have to be managed. They are all time-consuming and have costs, and there are many interactions between them. They form together a one-off exercise, namely a project, which lends itself to project management. This type of project will be more fully examined later in the book in the discussion of a case study concerned with the development of a new product, together with the design and construction of a completely new plant to manufacture it, for Photo Products Ltd.

Procurement means, and organisational environment

When a new capital facility is to be commissioned and established, one of the first things that has to be done is to decide on the means that will be used to achieve it. This stage, which may be thought of as being pre-project, does actually form a very important part of the project and has to be properly managed. There is an essential stage to be undertaken in relation to business decisions about the size, location and general purpose of the facility, backed by project appraisals both in terms of technical and economic feasibility. These interesting and necessary studies form part of the detail of Chapter 4 on scope, but for the purposes of the present discussion let us assume that we are concerned with a project where all such reviews have been completed and the scope of the project has been defined. It is likely that the means of procurement will be project-specific or at least industry-specific, and therefore it is appropriate to take a simple look at a number of cases. Given that we are concentrating on construction in this book, let us examine a building project first of all. The most common traditional practice was for the client company to appoint a consultant architect to help with the briefing of the project and then carry out preliminary designs for the purpose of obtaining rough cost estimates, and in order to let the client see what form the building would take. After a number of iterations of the design and costing, a decision to proceed would be taken and more detailed work would be put in hand. (This process is by no means as simple as is implied by this statement, and it will be further discussed in Chapter 4.) Having got to the stage of an agreed outline design for the project, a number of decisions have to be taken about the contractual method to be used in carrying out the detailed design and building construction.

In most buildings it will be necessary to employ specialist designers for the structural engineering and services, e.g. heating and ventilating, and it has to be decided whether these should be engaged by the client directly or by the architect. (The detail of this is included later in Chapter 11 dealing with forms of contract.) Mention of contract does however bring up one of the main decisions to be taken at the start of a project, namely the contract form to be used. At this stage of the book it is only necessary to point out that the contract to erect the building can take any one of a number of forms. Broadly contracts can be either lump sum where the contractor quotes a single fixed price for the work, or re-measured in which case the contractor's total contract figure will be based on the actual work completed, charged for at pre-agreed rates for each item of work. Within these two general forms there are many other factors to be considered, e.g. will the contract be based on fixed rates or will some measure of inflation be allowed for? Other forms of contract may include the detailed design stage, i.e. 'design and build', or 'turnkey'. Within the construction industry it is likely that civil engineering projects will be handled by engineers rather than architects, and there are correspondingly special contract forms for such work, e.g. the New Engineering Contract. The whole subject area of forms of contract is a complex one in construction, and has become a special interest of the quantity surveying profession, although this is a specialist group which does not exist in every country of the world. In the UK the quantity surveyors have been among the first professions to take up project management, and given the complexity of many contract forms this is understandable. It is of interest to note that some of the more recently developed procedures have been prepared by project managers who are themselves engineers (e.g. the *New Engineering Contract* and the British Property Federation System).

Some construction clients may themselves be big corporations who include among their staff building professionals, and as a result they do not need to engage specialist consultants. There may still have to be decisions taken about the form of contract, and whether the work will be undertaken by the corporation itself with its own regular or hired-in labour. Whatever method of procurement is chosen for a project, it is easy to see that the early stage of a project where decisions on these matters are being made calls for careful and effective management. If mistakes are made in setting up contracts, these can be very expensive to overcome, whether in terms of cost, time, or quality.

Installation of a management information system

This case of a non-construction project illustrates the same need for the means of procurement to be managed. A medium-sized manufacturing company with a long tradition and somewhat out-dated methods realises that it must establish a good management information system if it is to operate effectively and efficiently in future. Perhaps the first step in the project would be to engage consultants to study the business and analyse its needs. This would correspond to the briefing stage of a building project. It is likely that this would produce a report proposing a detailed study of the company's procedures and its information needs, followed by a systems design to provide for the means of meeting those needs. The systems design should be in terms of the functions to be performed, and the outputs required, rather than a specific software and hardware package. It is even possible (but unlikely) that a solution without computers could be proposed, and perhaps this is one of the main project decisions which has to be made. Another major decision is whether an existing 'off-the-shelf' computer package of software and hardware should be bought and customised to the company's needs, or whether client-specific software should be written; it is unlikely that special hardware would be designed, except for special attachments to production machinery. This decision can be a crucial one, partly in terms of cost but more importantly in terms of delivery date and reliability. Many computer projects have run into real difficulty as a result of this decision, either by trying in vain to modify an existing program which was not suitable, or conversely by grossly underestimating the difficulty of the task of writing client-specific software, and the time and resources needed. Some of the most notorious failures in information systems have revolved around this problem, again showing the need to manage the process properly, for example by pursuing more than one route for at least part of the way if the consequence of failure is dramatic.

Other types of project will have their own procurement decisions to be made. For example setting up a new production facility for a new product will have two inter-related concurrent projects, namely the new product itself and the establishment of new plant for its manufacture. The product design may well be undertaken in-house, but may also involve some input from a design company. Who is going to undertake the market survey, and how will this be fed back into the design of the product? With regard to the establishment of the production line, will this be done by modifying an existing line which currently produces an obsolescent product, and if so will there be a gap in production between the old and the new? Is this likely to be a problem, or can it be overcome by stock-building? Who will make decisions about the selection and purchase of new machinery? Is it feasible to modify the existing production line, and who will make this decision? Does the whole process of up-dating the product and plant require specialist advice, or will the whole project be handed over to another company? There are clearly many questions to be asked about the organisation of the changeover project, and these should be resolved at the outset. There are many projects of this type which have run badly, due to the lack of proper definition of the organisational structure; who is producing what, who is making decisions, and who is carrying responsibility. Again there is a need for the proper management of the means of achieving the project on time, within budget, and up to the specified performance. In some ways it is easier to measure the performance of this type of project, since this may be expressed in terms of a required output of product measured in units per week. In the case of a building project some of the performance criteria are long-term, e.g. withstanding gale-force winds which may not be experienced for many years, or the intervals between exterior redecoration. Similar discussions will take place on many different types of project, e.g. the development of a specialised hospital department (Manton et al., 1991).

Planning and progress

It was said in Chapter 2 that the management of any production process could be represented by the plan–measure–control cycle, and this applies equally well to project management. In the case of a project however, its unique nature makes planning both more difficult and more necessary. In the case of steady-state production the production plan may consist of little more than what was done last week. Even where circumstances change, there is usually some good recent experience upon which a plan can be based. In the case of a one-off project there is little or no direct prior experience upon which the planner can work. At the same time the fact that most projects are made up of a large number of interdependent items of work makes it all the more important that reliable plans are made; it is not very efficient just to let the project run ahead unplanned, in the way that a production line may sometimes be allowed to continue, on the assumption that 'if it was OK last week then it is OK for this week'. There is the added complication that where the

35

nature of the work on a project changes from week to week then different resources will be required. In fact one of the main differences between production and project work is that in the former the resources may be more or less static, whereas in the latter the resources needed change very frequently. In order to cope with this variation in demands for resources it is essential that the work is carefully planned. It cannot be too strongly emphasised that careful and detailed planning is one of the essential and most important aspects of successful project management. The realisation of this and the absence of appropriate techniques in the past led to the development of network planning methods, which form the main content of Chapter 5. Unfortunately there was a feeling in the early days of project management that networks would themselves solve all the problems of controlling projects, but this has proved to be grossly over-optimistic. It is now realised that network methods provide very useful tools, but as in all other activities they must be properly used and backed with a range of other methods. Before studying these methods in detail it is useful to see how the progress on a project can be measured.

Models in the measurement of progress

Measurement of the output from a steady-state production process is usually fairly easy by counting or measuring the output in terms of units per week. There is a difficulty in the case of project work because the actual nature of the work will often change on a day-to-day basis, and furthermore the output does not lend itself to being counted in terms of uniform units of production. Over a period of years a number of ways of measuring project progress have been tried, and most of them have been found to have short-comings. It is useful to describe some of these in order to understand why networks have provided a good basis both for planning and for measuring progress. One way of doing this is to consider a simple example which is not specific to any one industry or type of project.

Here we consider only the implementation stage of a project, assume that all the preliminary decisions and design have been completed, and an order has been placed for the work to be done. The value of the project is £1 million and the planned project duration is ten months. There is a strong temptation to measure progress in financial terms, partly because one of the main parameters of the project is its value, and partly because most organisations will have most of their records in terms of money, e.g. wage records, purchases of materials, hire charges for equipment, and receipts from clients. This may have led to the adoption of the very simplistic yet widely-used basis of measurement on a project which assumed that the total project cost could be divided by the duration of the project to give a notional value of work per month which must be completed. In the example quoted above this would mean that the total cost of £1 million would be spread over the ten-month period to give a notional value of £100 000 per month for the work to be done. This is illustrated in Fig. 3 which shows the cumulative cost of the work over the period of the project. The planned cost is shown as a straight line rising uniformly over the ten-month period. The actual cost of the work achieved can then be plotted on the same basis and a comparison made between work planned and work completed, as measured by the costs incurred to date.

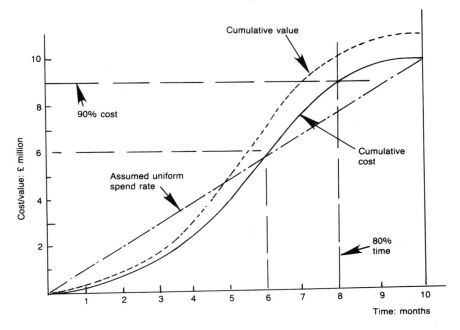

Fig. 3. S-curve showing cumulative cost and value

There are two problems here. It is often the actual costs of work done rather than a true measure of what has been achieved that is recorded, simply because cost information is readily available. The rate of activity on a project is seldom uniform. The use of cost data can be misleading, because they will include the cost of any abortive work and the correction of faults etc.; what should really be measured is the real value of acceptable work that has been completed (in the case of a building contract this would be based on an interim valuation of work).

A build-up of the rate of working presents the second difficulty, and requires more detailed explanation. On any project it is likely that initially there will be relatively few people working on it, but that the team will build up to a maximum number and then decline towards the end of the work. (For a computer installation this would mean a few systems analysts at the start, building up to hardware specialists, programmers, applications writers, and installation teams, and than tailing off at the end to the small commissioning team. In the case of a building the initial small team would be the excavator operators, then concrete gangs, bricklayers, carpenters, electricians, plumbers, with probably only the painters and cleaners left at the end.) It is not only the direct labour costs which follow this pattern, since it is likely that the quantities of materials used and equipment hired will roughly follow the number of people working on the project, indicated by the lines on Fig. 4. This well-understood phenomenon is represented by what is known as an 'S-curve' when plotted as a cumulative figure, as shown on Fig. 3 above and further discussed in Chapter 6. It is when we compare plan with performance that the importance of the two lines on Fig. 3 can be seen.

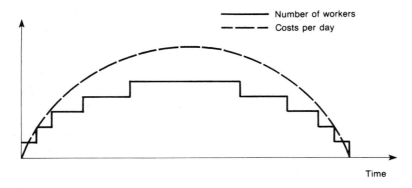

Fig. 4. Spend rate and labour strength during project

Cumulative-cost curves are applied in the simple example discussed above, and Fig. 3 shows both the uniform cumulative curve (straight line) and the non-uniform line (S-curve). Consider initially that we have used the straight line as the planned rate of work, but that the actual rate of work has followed the S-curve, and assume that the project has been completed on time. There will be three points at which the two lines coincide, at the start and finish of the project and somewhere around its mid-point. At each of these points a check on the diagram will indicate that the project is 'on programme', but let us consider more closely the eight-month stage of the project (i.e. 80% of the duration) where the straight-line target will indicate a project value at that time of £800 000 (i.e. 80% of the project total value). Examination of the S-curve will however show that the project value at that time ought to be around £900 000 if the work is to be completed on schedule. If in reality the project value at the eight-month stage is only £800 000 then it can be seen that the project is behind schedule and is following a cumulative value curve which will overrun by one or two months. This of course is what happens in practice; many projects seem to be going well for much of their duration, and then for some unknown reason near the end are found to be behind programme. There are many reasons for this occurrence among them being the inherent optimism of traditional managers (see Chapter 10), but also the fact that any activities which are not urgent tend to get left 'until there is more time', but then bunch up at the end and run into resource problems (see Chapter 9). The conclusion from this simple fictitious case so far is that the measurement of cost or value against a straight-line assumption is not a sound basis for the measurement of progress on a project.

The concept of an S-curve is useful in terms of project financial control but it does have limitations as a means of measuring actual progress. This can be illustrated by reference to another simple case, that of a new product which will be packaged in a totally new way, and distributed through a new dealer network. The project has many strands to it, and it may be that the new packaging and distribution systems make very good progress and create real value for the project. They will, however, have done nothing to advance the design and manufacture of the product itself which may well be falling behind programme. In a similar situation a building project may appear to be going well because its first few months' value is up to the S-curve target; closer inspection may reveal that much of the value created has been due to the construction of a high perimeter wall around the whole site, valuable in itself but doing nothing to advance the actual construction of the building. The final conclusion of this discussion of monitoring progress is that money does not provide a suitable unit of measurement, except in the sense of controlling finance. Other industries do not use money as their main measure of output, the oil industry produces barrels of oil, the motor industry counts numbers of vehicles, coal-mines produce tonnes of coal, and service industries such as transport record passenger-miles. One of the challenges to project management has been to devise a means of measuring progress when there is no consistent unit of output such as tonnes or passenger-miles, and which does not use money as its base. The method adopted has been to create a model of what has to be done and then compare actual performance with the model. The model usually takes the form of a network as this can represent very well the interactions between all the activities which make up a project. The models exist either as drawings or as a series of linked statements in a computer. They serve a number of purposes, starting with providing a means of thinking about the project, then acting as the basis of a plan to check time, cost and feasibility, and finally acting as the basis for comparing actual progress with the plan. Models in many situations are also used for experimentation where actual testing is either too difficult, expensive or dangerous (e.g. the ultimate load testing of a bridge). Networks as models permit the testing of activity sequences when clearly actual tests would be virtually impossible in one-off project work. The foregoing discussion has led to the arguments leading to the development of networks in project planning and control. The detail of how this is done is the subject of Chapter 5.

Time management

Much of the previous section covering the management of planning and progress is concerned with the management of time, and indeed the time element is a predominant consideration in networks. At the start of this section it may seem somewhat simplistic to ask what we mean by time, but it is useful because there are at least two aspects which must be considered. There are effectively two types of time in common use in projects, and there is confusion between them. First there is time in the sense of resource use, e.g. the worker-hour. There is a long tradition in many industries, especially heavy mechanical engineering, to measure work in terms of man-hours. This is because it permits the planning of the use of resources which may be fixed in total, and also is closely related to costs, especially labour costs which form a fairly high proportion of total costs in the labour-intensive project industries. The man-hour is a useful means of stating the work content of an activity, often a truer work measure than the cost in money terms since it is not subject to inflation and does not carry overhead or profit elements; it does nonetheless suffer from some variability due to changes in working conditions and working methods. It is sufficiently useful as a base measure for many operations in construction and heavy engineering to justify the publication of 'standard' man-hours for a wide range of tasks.

The other use of time is in the sense of elapsed time or clock time, and this has to be distinguished from work content. The difference can perhaps be best illustrated by an example. A particular operation in the execution of a systems analysis of a computer application is estimated to require 30 man-days of work by a systems engineer; this means that it could be completed by 1 person in 30 working days or in 15 working days by 2 people. In the first case the elapsed time would be 30 days and in the second case only 15 days; the work content would be 30 man-days in each case. It may seem tedious to emphasise this point but the author has seen many instances of confusion between the two uses of time. One of the topics covered in Chapter 8 concerns the ways in which we may seek to reduce the elapsed time on a project, but for the present it is worth giving it some attention. The above simple illustration indicated that it might be possible to reduce the elapsed time required for an activity by doubling the number of people working on it. Depending upon the nature of the work it may or may not be possible to make a further similar reduction by allocating another person to the task, making 3 in all, in an attempt to reduce the duration to 10 days. This cannot go on for ever and it would probably be quite impossible to seek to complete the work in 1 day by using 30 people. It may well be the case that even with 3 people the work cannot be totally shared out and that it would take say 12 days of elapsed time (rather than 10) to complete the activity. This of course means that the actual work expended on the task would have increased from the original of 30 man-days to 36 man-days, with the consequential increase in cost. An alternative to increasing the number of people working on a task from 2 to 3 would have been to ask the 2 people to work overtime by increasing their working week from 40 hours to 60. This might just succeed in finishing the task within an elapsed time of 10 days and a work content of 30 (standard) man-days. A clear understanding of the two uses of time is essential for anyone taking up project management, especially if they come from a background of contract estimating in which all work has been evaluated in terms of man-hours. In most projects the client will primarily be interested in the duration or elapsed time of a project, or perhaps even more truthfully in its completion date. While the supplier or contractor is also interested in completion by the agreed date, there is no doubt that the man-hour content of achieving that date is often of equal importance.

The preparation of a working plan may be undertaken on any of the many ways of planning and measuring work on a project; these are primarily based on time, and several of them are looked at in detail in Chapter 5. One of the traditional ways has been to represent separate tasks in the form of a bar chart, but as will be shown this does have some shortcomings. The same also applies to a simple list of target dates. It was because of these that network methods were evolved, and their use is now widespread in project work. It must be pointed out at this stage however that networks do not provide all the answers, and there are many instances where barcharts and/or lists of dates are the most appropriate. Time planning and control are essential to good project management, and it is useful to compare the planning of a project by analogy with its design. The technical design of a new structure will involve many sequential stages.

(*a*) The initial concept.

(*b*) The briefing of its form and specified performance.

(*c*) The loading and other requirements.

(*d*) Stress analysis, other calculations and sizing of members.

(*e*) Consideration of alternative designs and making choices.

All of these will be completed before the final stage:

(*f*) Preparation of the working drawings.

The planning of the erection of the structure is in effect the design of the erection method, and will go through a corresponding series of numbered steps, as follows.

(*a*) Analysis of what has to be done, in terms of activities.

(*b*) The logic sequence of the work.

(*c*) The elapsed time and resource needs of each activity.

(*d*) The calculations of project duration and its adjustment.

(*e*) Consideration of alternative methods and making choices.

All of these will be completed before the final stage:

(*f*) The preparation of the plan of action for completing the work.

In much the same way that it is not normal practice to go directly to the detailed working drawings at the outset of the design stage, it is not good practice to go straight to the barchart as the plan of action for the construction. Some planners attempt this and usually find it unsuccessful. The underlying theme in both design and planning must be.

(*a*) Define what is to be achieved.

(*b*) Analysis and understanding.

(*c*) Data input.

(*d*) Detailed calculations for achieving the target.

(*e*) Comparing alternatives and choosing the most appropriate.

All of these must be completed before the final stage:

(*f*) The stated plan of what is to be done and how.

The final plan of action for the implementation of a project may exist in the medium of drawings, written statements or computer outputs. It may take the form of networks, barcharts, date lists or even written words. Barcharts have the positive advantage of giving a pictorial representation of progress to a time scale and are readily understood. They are less useful at the earlier analytical stage, for which some form of network is more appropriate, and is really essential. The essence of this could perhaps be stated: networks for analysis, barcharts for presentation. This is not a golden rule, but is a guide to good practice.

Cost management

The traditional way of controlling costs in industry is to set a budget forecast and then compare actual performance with the budget. It is then usual to highlight any significant variances from budget, seek the cause and take remedial action where appropriate. This approach is entirely consistent with the plan–measure–control cycle previously described in Chapter 2 (Fig. 2.). There are often two broad approaches to cost control; one is more global, i.e. to plan and then measure the actual inputs to the process being considered, for example the consumption of raw materials, utilisation of labour, use of equipment, energy consumption etc. This approach will cover total use of an input item rather than cost per unit of output, but it does give good overall control and can quickly alert management to major problem areas such as power or other energy leaks, wastage or theft of materials or components, various types of fraud and so on. This type of control is unlikely to pick up smaller but still important discrepancies in terms of efficient use of materials, operator time loss because of lack of parts etc. One of the ways to track this type of problem is to use the second approach to cost control, a system of 'unit costs' by which all the cost inputs to a process are recorded on a week-by-week basis and then any trends can be identified and acted upon. Such unit costs would be separated into a number of categories including main material inputs, labour, equipment use, energy and so on. An extension of this procedure is not simply to compare one week's cost figures with those of the preceding weeks, but to establish a 'standard cost' which is what the unit costs should be if everything goes well. The object of this is to avoid figures which are consistently poor and therefore do not show up in comparisons. Regular checking against a calculated standard will highlight such continuously poor performance. One of the problems with the use of the standard cost method is the difficulty in establishing the initial figure for inputs, and this complex task is the responsibility of work study personnel. It can be fraught with difficulties especially where it is linked to the setting up of targets for bonus schemes, and can itself be a costly exercise. It is through this approach however that detailed studies of the production methods in use can make significant improvements by examining carefully all aspects of a process, e.g. the physical layout of the workplace, the storage of components, the use of working jigs in place of measurements, the automation of simple repetitive procedures, and many other similar items.

The management of costs in projects is difficult. The above description of cost management in steady-state production is given not as a means of managing costs in projects, but to show that there is a contrast between the two types of industry in the way they carry out cost management; it is necessary that this difference is appreciated if good control of costs is to be exerted. There has been much time and effort wasted on projects trying to undertake costing exercises on the principles outlined above which were designed for steady-state production. As an example the author was engaged in a study to record and compare the weekly costs of fixing and striking shutters for a large concrete wall; over a period of several weeks apparent variations in cost were found to be due to inconsistencies in recording and major changes in weather conditions. These disturbances took some weeks to resolve, by which time the wall was complete, and the only lesson to be learned was the futility of the exercise on a one-off job. Subsequently work was done on other sites, and from this some general principles for the design and use of shutters were established, but it did not produce a good weekly cost control system. The unique nature of projects and the lack of continuity of any individual item of work does mean that the 'conventional' approach to cost control is not really applicable. As a consequence a range of other techniques have been evolved to manage costs, and these are described in Chapter 6. Significant work has been done in this area by J.R. Turner, as detailed in *The Handbook of Project-Based Management*.

Quality management

This simple two-word phrase has recently become one of the new management areas of great interest and the subject of many books, articles, seminars and courses. It has been brought about by competition among manufacturers who now seek to expand their markets by improving quality rather than simply competing on price. The electronics and motor vehicle industries have made great advances in terms of quality, and there are now major efforts being made to bring the benefits of good reliability and quality into the project-based industries. Quality is the subject of Chapter 7 where again the special case of one-off project work is described. The most important point to make here is that the slogan often used in quality management 'get it right first time' is nowhere more true than in projects, simply because there is seldom a second time. In manufacturing it may be desirable to get things right first time, but it is usually possible to throw away the first unit of production even though it may cost money to do so. This is a luxury which is not possible in project work.

A fundamental point in the consideration of quality is the definition of what we mean by the word, and it is important to make this clear. There are at least two aspects which we must think of separately. In common usage we think of high or lower quality goods when what we really mean is their standard. For example, we may think of a Rolls Royce car as being of high quality compared with the lower quality of a mass-produced car such as a Ford or VW, when what we really mean is that the standard of specification is higher. The meaning of quality in production terms is the degree to which every unit of production meets the standards which have been set. In relation to motor vehicles, customers have every right to expect that the Ford or VW vehicle will meet the stated standards and will be reliable, and in fact these and other motor manufacturers have been able to achieve very good reliability rates. It could well be argued that in mass-produced vehicle production it is easier to maintain high standards because so many of the processes are automated and therefore not subject to the risk of human error that exists with the hand-built car. This does provide a link with project quality, since the production of individually built vehicles is very close to being project work where a rather more meticulous scrutiny of every operation becomes necessary. The distinction between the two aspects of quality that we have identified is that the first can be thought of as 'quality decision', and the second as 'quality attainment'. In this text the phrase 'quality assurance' has been avoided since it has a very specific meaning in some contexts.

Another important area concerning quality in projects is the interaction with time and cost and this is pursued in Chapter 8. There is often a trade-off between the specified standard for a product and its cost, either in relation to the materials used, the allowable tolerances, or other attributes. There is sometimes also a trade-off with the time required, especially in the case of projects. This is particularly true when some slightly defective work is detected and a decision has to be taken to correct it at the expense of a delay in the project. This is often a 'catch-22' situation to which there is no easy answer, but one which does need to be addressed, as is done in Chapter 8.

The management of people

It is in this area that project management probably has most in common with the management of other types of activity. Clearly any undertaking that employs a significant number of people will have to use a range of skills concerning the management of people, including communication, leadership, delegation, recruitment, selection and training; there are many good texts on these and it is not relevant to discuss them at length here. There are aspects however where project management does emphasise particular aspects of people management, and these are worth clarifying. Chapter 10 is entitled 'People in project management', but is really mostly concerned with the characteristics and skills required to be a project manager, and the ways in which teamwork plays a major part in project management. Our immediate concern here is the management of other people in the project, and this includes not only the operatives executing the work but also the client's staff, professional consultants, government officials, bank representatives and so on. In Chapter 2, the discussion of jobbing work pointed out that much of the influence over what is done in a project is literally in the hands of the operative or individual worker. Furthermore the work that has to be done will change from day to day, and it cannot be assumed that just because today's work is going well that tomorrow's will also be carried out properly. A simple and perhaps trivial example of this is the case where a brickwork gang on Monday starts the building of a house wall; by Tuesday it is clear that they are careful workers with all courses laid truly to line and level, proper bonding observed, all joints clean, very little waste, and with good progress being made. The foreman is confident that they are reliable and is too busy to look at their work again until Thursday, by which time the wall has been well built, but with window openings in the wrong place! How often are errors like this found on a building project and what does that tell us about the need for close control in projects, with detailed supervision of work? More will be said about this in Chapter 7, but it is worth stating here that the consequences of such errors in projects can be very great as the cost of correcting or re-working can exceed the initial cost of the work in both money and time. There is a great need for clear instructions in delegation, good supervision and appropriate training. This contrasts with steady-state production where there is still a need for supervision but the same instructions will apply for several days at a time, and the only real management task is to ensure that things keep running smoothly. The management of cost, time, and people is further discussed by Wearne (1989).

It is not only in connection with operatives that close supervision in projects is needed. For example it is common in many project-based industries to employ specialist consultants or sub-contractors to carry out part of the work. This again requires a special approach. For example, a manufacturing company may sub-contract some of the components of its product and can call upon the supplier to submit samples for checking; it then has the ability to reject these samples or any subsequent items which do not come up to the required standard. In the case of project work this may not be so easy, as the sub-contract work is being done on the project site, and in real project time. For example, a specialist machine-tool company may be installing a new production line and chooses to sub-contract some minor building works to accommodate the new equipment. The fact that the building works are small in content and value does not reduce their importance in getting the project complete; a small roof leak which does not become apparent for a few weeks, by which time delicate plant has been installed, can prove to be very serious. Most project managers will be able to relate similar horror stories.

Project managers will also have to learn how to deal with designers and other professionals who frequently regard themselves as the most important people in a project and therefore have to be handled very carefully. Even more difficult is the client's representative who may be extremely demanding and will often point out who is paying for the project and therefore who has the right to call the tune. It does not always go down well with clients to point out the terms of appointment of the project manager and/or the terms of the contract, even though these may give the project manager complete authority over the matter in question. This type of people management requires great diplomacy as well as good communication skill.

Risk management

This is another of the phrases which has come into the vocabulary of management literature in recent years, and like so many of these concepts means different things to different people. For example, the insurance industry sees itself as the home of risk management, but while insurance makes a major contribution to the subject there are many other aspects. Essentially risks arise because there is an element of uncertainty in everything we do, and also a variability in the outcome of every process. What we are seeking to do in risk management is to manage variation and uncertainty. The actual concept of risk is a measure of the consequences of the occurrence of the unexpected, rather than the chance or probability that it will happen. There are many risks associated with project work, only some of which can be covered by insurance. On a building site it may be possible to insure against flood or severe wind damage, but it is unlikely that compensation could be obtained for the slowing down of work due to high winds or prolonged rain; these are not disasters but merely inconveniences. There are commercial risks to be considered in relation to the prices of components and raw materials, uncertainties about labour rates that will prevail, the possibility of new legislation which will impact on the project, whether the client will want modifications but is not prepared to accept a delay, and many others which are examined in Chapter 13, and are discussed in an introduction to the subject of risk management (Thompson and Perry, 1992).

Project success/failure

At the end of a project it should be possible to make an assessment of the extent to which it has been a success or failure. In order to do this it is helpful if criteria are established at the planning stage which can be used as the basis of the assessment. Some of these criteria are easily identified.

- Was the project completed by the target date?
- Was the project completed within the budget?
- Has the completed project met all the performance standards set?

Answers to these questions will to some extent be dependent upon the point of view, whether it is that of the client, designer, contractor/supplier, or the actual user. Most parties to the project will be interested in completion on time, but the attitude to cost will vary. The client will be keen to keep the total cost to a minimum, but the contractor/supplier will wish to maximise the profit margin between actual cost of work and the price paid by the client. The designer will wish to see a good job well done, and both the client and the final user/occupier will be anxious to see that quality and performance are as specified.

Some projects will fail to meet their expectations in some respects, and some may be total failures, even to the point of being abandoned before completion. Generally all parties will celebrate success, and are not keen to publicise failure, but it is useful to undertake an analysis of project performance, so that lessons may be learned for the future. Major projects and their shortcomings are likely to be discussed publicly.

Facilities management

This subject is not so much an essential part of project management but rather an add-on optional extra, and indeed many project managers will never have any contact with it. The subject has been studied for some time, and previously was referred to as asset management, but this caused confusion with the financial operators who simply saw assets as locations to invest money with a view to obtaining reward, either in terms of dividends or capital appreciation. Asset management was really about making the best use of buildings, plant and equipment, i.e. assets which were in the possession of the organisation for the purposes of their production. This could mean the best utilisation of a fleet of aircraft, the optimal production schedule for a factory, or the most effective use of building space in a hospital or university. Asset management has now come to be known as facilities management and has become part of project management partly because project managers are the people who install the new facility or modify the old one. It is common practice for large organisations to have a plant strategy or estate strategy to create, maintain and operate their stock of plant and buildings; the preparation of this strategy and its implementation and updating is often made the responsibility of project managers within the organisation.

DISCUSSION OF PROJECT MANAGEMENT

This brings us to the end of the general discussion of the subject of project management and the ways in which it compares with and differs from the management of manufacturing. This has been given at some length in order to give a good understanding of the fact that we are dealing with a different subject area and have at times to think differently. The following chapters deal with the 'how-to' of project management, namely the techniques and methods which have been developed to help practise the discipline. Further reading on the general applications of project management can be found in various books (Cleland and King, 1988; Reiss, 1992; Stallworthy and Kharbanda, 1984). A number of major projects which give a most interesting context to the subject are also described (Morris and Hough, 1987; Leech and Turner, 1990; Smith (ed.), 1995). There is a thorough and perceptive history of project management, illustrated by reference to many well-known projects by Morris (1994).

Terminology of project management

Throughout this book and most of the relevant literature there are many specific words and phrases which are used to denote precise concepts. Where it is important to be accurate these terms have been defined, but in addition a language of project management has evolved and is explained (Cleland and Kerzner, 1985), and is also set out in a number of British Standards (BS 4335, BS 6046 and BS 6079).

BRITISH STANDARD BS 6079: 1996, Guide to Project Management

BS 6079 was published at a time when this book was in the early stages of printing. It is therefore only possible to give some brief discussion of the document itself, rather than comment upon the use to which it has been put in industry. First impressions however would indicate that if its general messages are taken to heart by clients and contractors/suppliers then it will indeed improve the performance of projects. This is its stated objective, since it starts from the position that 'the overall track record of British organisations in managing projects, including takeovers, leaves much to be desired'. This is a view which can be disputed however, since within recent years the performance on many projects has been very good, especially where project management has been properly applied. There are still several much-publicised failures, many of these due to the fact that thorough project management has not been put in place, especially by clients. BS 6079 addresses this point and its stated objective is 'to help managements to deal with these situations more efficiently and effectively'.

The new British Standard was prepared by a committee, many of whose members are related to building and civil engineering or other heavy construction, but also including aviation, electronics, telecommunications, defence, quality and others. Participating organisations are listed alphabetically, with the APM first; reading the document also indicates that the influence of the APM has been a very important and leading one. The introduction to BS 6079 gives a good one-page statement of the development of project management and its place in modern industry. It is also clear that it provides a set of guidelines for newcomers to the management of projects and an *aide memoire* for those with some experience. It is not a prescriptive standard which lays down mandatory requirements, which was a fear that had been expressed by a few people before it was published.

There is a useful list of definitions of the terms used in project management. All the terms used in this book and most of those used in other texts are consistent with BS 6079 which also includes some additional definitions which are of note. One of these is 'task' which is defined as 'the smallest indivisible part of an activity when it is broken down to a level best understood and performed by a specific person', although this does seem to be too restrictive; in many projects there may be tasks which have to be performed by a team. Another definition is 'statement of work' (SOW), which is 'a document stating the requirements for a given project task'. Given that the task is stated to be the smallest indivisible part of an activity there will be a very large number of tasks and consequently a large number of SOWs. This could lead to a great deal of paperwork which might become a distraction from the real job of getting on with the project activities.

The content of BS 6079 is a very good general statement of project management, and it provides a list of project phases very similar to the one given in Fig. 5 (page 52) in the next chapter of this book, including the frequently ignored final phase of 'disposing of residual assets (liabilities)'. The standard also deals briefly with the important area of project organisation. Considerable attention is given to the project management processes of planning, measurement and control; this includes work breakdown structure, task dependency and criticality, costs, earned value, cash flow, deliverables, milestones, resources, quality, configuration management, risks, procurement, subcontracting, project personnel and their required skills, project life cycle and several other related topics.

Overall BS 6079 is a good concise statement of what has to be done, rather than how to do it. In under 50 pages it gives a broad introduction to how the management of projects should be approached, and as such it will provide useful guidance to the senior management of both clients and contractors/suppliers who have not previously made proper use of project management; it is not a superficial document. At the operating level these guidelines remain relevant but will need to be supplemented by much more detail.

4 The management of scope

THE DETERMINATION OF SCOPE

Chapter 1 was concerned mostly with a discussion of what we mean by projects and project management, outlining the characteristics of one-off undertakings and how they can be planned and managed. It looked further into projects themselves, and sought to differentiate between hard, soft and open projects, and broadly these can be thought of as follows

(a) hard projects involve hardware
(b) soft projects are information and decision-based and exist mostly on paper
(c) open projects are on-going activities which do not meet all the normal criteria of project management and are therefore set aside from the discussion of this present chapter.

The immediate purpose now is to set down what it is that we are seeking to manage when we take on a particular project. The same approach can be applied in general terms to both hard and soft projects, such as the case of a new product in manufacturing, the acquisition of a new company, the design and construction of a new building, or the writing of a new procedures manual for a quality management system. It can also be applied to 'management by projects', which is the use of the methods of project management within an organisation which is not normally thought of as being project-based. An example of this is where a manager is responsible for the preparation of the annual report, the introduction of a staff development system, and many other similar tasks which can be treated as a series of projects.

The determination and definition of scope of a particular project is one of the most difficult but also most important stages in project management. Basically it is answering the questions 'What is this project we are about to undertake? What is included and what is excluded? What standard of performance is required?'. There are clearly no simple answers to these questions, and immediately we see a number of other questions which may have to be answered first.

Who is the client or sponsor? This is frequently a matter of determining who is paying, but this in itself is not always immediately clear; in many cases the payment may be coming from a combination of sources, the client company, its regular bankers, a merchant bank and so on. These can all have an influence on the way the project is managed.

Who is to be the user of the finished project? In many cases the client organisation may not be the actual user, as in the case of an office property development where the formal client may be an insurance company or a pension fund, and yet the user will be a tenant with no formal or lasting link with the owner other than a lease of occupation of the building. Even on a much smaller scale the introduction of a new production line within a factory may be managed throughout the project phase by the development department of the company, but will later be operated by the production department.

What is the economic, social and legal environments of the project, and will these have an impact on the way that it is defined? Are these environments sophisticated or simple? i.e. highly industrialised and organised, or remote? Are they stable and predictable, or subject to rapid variation which would have an impact on the project?

Who is to make the major decisions? It is likely that the client organisation will make the first decision, namely that a project should be considered in the first place. It may be that some projects are thought up by entrepreneurs who then go and seek out a potential client, but even in these cases a major client decision is involved. Perhaps the most important and first decision by the client is how the project is to be implemented, and who will be involved in its briefing. Does the client feel competent to undertake this task alone or will professional help be required? The decision process is an important part of the scope definition — the various routes for implementation are set out in Chapter 9.

The importance of defining owner and user

An interesting example of a project where the definition of owner and user have become critical is the Channel Tunnel. The formal owner is Eurotunnel, which is not a company operating in any other area but is a project-specific undertaking. The financial support for the project has been partly through shareholders, with the support of no less than 225 banks. The user of the tunnel could be any one of a number of train operators, the main ones initially being Eurostar and Le Shuttle. The actual construction was completed by Transmanche Link, a consortium of companies set up for this one project. While there were a few technical problems the actual construction work was well executed, but there have been major financial difficulties which have become apparent some time after the tunnel was put into regular operation. Resolution of these problems will take a long time, and will involve not only the owners, bankers, users and constructors, but also the French and UK governments. The last-named are significant here in the discussion of project scope, because at the outset of the work it was made clear by the two governments that the tunnel would be a privately funded enterprise. It was however the UK and French governments that were really the project sponsors, setting up the organisational structure that would be needed to fund, build and operate the tunnel. It was also the governments who were in control at the time of the railway systems as a whole, and also had powers to influence the competing cross-channel transport systems. It is therefore not surprising that when in financial difficulty Eurotunnel turned to the governments, as well as to its commercial partners for support. With the benefit of hindsight, which is a common and very valuable approach in post-project review, it can be pointed out that in the definition of scope it was not enough to think in terms of the tunnel and train operations alone, nor even of the links between the tunnel and London, Paris and Brussels. For a project of such magnitude the whole transport system of Britain and mainland Europe is a major component of the social and economic environment. While it was obviously discussed at the outset it is not clear that such discussion was part of the defined project scope and it is now obvious that such major international projects must be seen in a very wide context.

Establishing scope — case studies

While reference to famous major projects is interesting, most project managers will have to contend with much smaller and less well-known undertakings; many of the principles remain the same however. Rather than generalising too much about all projects, this chapter will deal with a few cases of particular industrial settings. Since this book is primarily intended for the construction industry, the first case is that of a building project. A case from a different industry (photographic equipment) follows later in this chapter. There are many different ways of going about the organisation of a building project using a range of forms of contract and many of these are compared in Chapter 9; the following discussion does not make reference to all of these but sets out a general framework based on the conventional competitive bid structure, with occasional reference to variations which are sometimes used.

Typical building project scope

Fig. 5 shows the possible different stages of a generalised building project. Any one particular project will include many but usually not all of these steps, but again it is essential that a clear decision is taken at the outset to define which are included and which are excluded, and where the responsibility for each lies.

The concept of the project and its purpose This is likely to emanate from the client, but may also be generated by central or local government. It is the first gleam of an idea which may need a great deal of refinement, possibly with the help of an architect or engineer.

Determination of the need This may well be a commercial or social exercise which will require an input from relevant specialists such as transport planners, health or education authorities and so on. It will cover such decisions as the location of the building, its general purpose, its size, and possibly the general standard to be specified.

Appointment of the design team This was stated above to be one of the most important decisions to be taken by the client, since it is from this point that the design and procurement will follow. In the case of important and prestigious projects this selection is sometimes carried out by means of an architectural design competition, but this does present a dilemma. The winning design may be judged on the basis of all the criteria set out in the paragraph below on selection, but it is likely

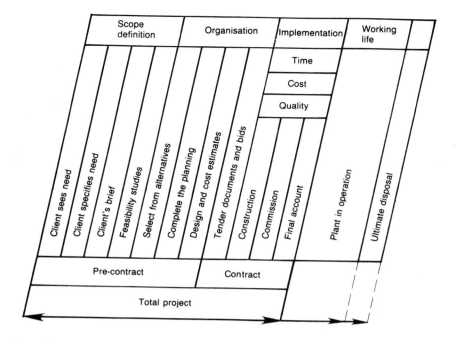

Fig. 5. Stages of generalised construction project and its full life

that visual appearance will attract much comment, especially from the press and public at large. The dilemma is however as follows; a competition may produce on any one project the best general design, but does not necessarily select the best architect in terms of capability in detailed design and implementation. The alternative approach of selecting an architect on the basis of reputation and past performance is likely to ensure good implementation, but may on the other hand not lead to the ideal design from the client's own viewpoint. Some of the more famous architects tend to dominate the conceptual design phase and offer the client relatively little opportunity to influence the design. It is probably fair to say that in engineering projects the client will have very little influence on the design which will be entrusted almost entirely to the specialist consultants, but may have produced a very tight brief within which the consultant has to work.

The client brief The brief will then be able to proceed, taking account of the purpose of the project and what facilities it is intended to provide. This stage of the project is often completed by the client and the architect and/or engineer working together. It is only in cases where the client company is set up to procure a series of buildings that it may have its own staff capable of carrying out the briefing without external help. This is an extremely important stage, as this is where many decisions have to be taken which have long-term influences which cannot readily be changed. It is important that adequate skill and time are devoted to the briefing as it is here that expensive errors or omissions can be made.

Feasibility studies and their appraisal   These usually form the first creative stage of the design process. In the case of a small project it is likely that only one design solution will be proposed, and provided that it is acceptable that solution will be refined to become the adopted design. On a large project it is possible that several different outlines are drawn up and comparisons made between them, in terms of cost, feasibility, functionality, durability and general appearance. Before moving on to the next stage of selecting one of the alternatives it is necessary to take the decision on whether or not to proceed at all, and this is likely to be based on economic and social appraisal criteria, as discussed in some detail in Chapter 12. In all projects there should be a clear point at which this decision is taken, and the project manager should always be prepared to abort the project at this stage; at any earlier point it is not likely that enough information is available, at any later time there will have been a commitment to fairly expensive development and design work. This may have the consequence that there is a great reluctance to terminate work, and a project which should never have been allowed to proceed runs on into disaster. The subject of project cost is largely covered in Chapter 6, where the question of cost escalation is considered, but one point is relevant in this discussion. If work is allowed to proceed on a project which is inherently too expensive there will be continuous pressure to cut costs, and this can only be done by lowering the standard of work specified or by reducing the scope by deleting 'non-essential' items. The acceptance of lower quality standards should never be contemplated, and non-essential items should never be there in the first place; it is therefore at the appraisal stage that a view should be taken on the economic viability of the project, and if it cannot be afforded then it should either be abandoned or totally re-designed.

Selection of the preferred alternative   Selection should then follow taking account of all the factors listed in the feasibility stage above. There is sometimes a temptation to give undue weight to the appearance of a building, as this is the one aspect that can be readily seen by the client who may not be skilled in visualising how a building will actually work, what it will feel like to live or work in, and what its operating and maintenance costs will be. The problems of this approach are sometimes compounded by the use of an architectural competition to select an architect for a project, a device which can have some undesirable outcomes, as discussed above.

Completion of the detailed design and target costs   Completion can proceed once the preferred alternative has been decided upon. This may be a fairly lengthy process but it should not be used as a time when shortcuts can be taken. One sure way of opening up the possibility of contractors' claims is to proceed with the tendering process before the design is properly worked out; this does not necessarily mean full working details but enough to define the work for pricing purposes. There are special forms of contract which seek to overlap the design and construction phases in a fast-track approach, and these are dealt with in Chapter 11. One of the purposes of completing the design is to provide the information for the calculation of reasonably accurate cost targets, usually worked out by quantity surveyors in the case of buildings, and by the design engineers in the case of heavy construction.

The preparation of tender documents   This will often be started whilst work on the design is under way, but the bill of quantities cannot be completed until some time after the design drawings and specifications are complete. All of these sets of documents are used as the basis of the pricing process by contractors, and subsequently become part of the contract documentation. Tenders may be sought either from a selected list of bidders, or in the case of open tenders from anyone who may be interested. These procedures are covered in more detail in Chapter 11, along with a description of other forms of contract, which may omit some of the documents listed here. It is usual to allow bidders a few weeks to make their estimates for the contract work and to submit a tender; it is not good practice to rush this stage of the process as it can give rise to subsequent problems if errors are made.

The evaluation of bids and award of contract   Evaluation will follow quickly, partly because all parties are keen to know the outcome and to get started on the physical work on site. It is necessary to ensure however that the selected bid is arithmetically correct and that all items have been included, because in many forms of contract the full list of items is regarded as the bid rather than the sum quoted at the end. If there is a discrepancy between the list and the sum then the former constitutes the bid. Errors do arise from time to time, especially if the tender period is too restricted, or if changes are made to the bill of quantities during the period between the issue of tender documents and the return of bids. The evaluation of bids is not simply a matter of checking arithmetic however, and it is necessary for the quantity surveyor, architect and engineer to satisfy themselves that all items have been reasonably priced without any undue loading of items where quantities may be in doubt. The selection of the winning bid may simply be based on the lowest price, and this is sometimes specified at the outset. There may be cases where there are reasons not to accept the lowest bid, and a decision is taken to select the second or some other bid, but care must be taken with this approach as is discussed more fully in Chapter 11. Once the winning bid has been selected it is usual to issue to the bidder some form of letter of acceptance, followed at a later date by a formal contract for the work. In some cases it may be found that there is a significant difference between the estimated costs worked out at the project appraisal stage and the selected bid price. If the bid is regarded as being too high by the client it may be necessary to reduce the scope of the work by the deletion of a number of items. This may well involve further discussions with the selected bidder, and if the changes are major it may be necessary to re-submit the bid. The client should not however see this as an opportunity to seek a reduction in the bid price.

The preparation of working drawings   Preparation may well have started during the earlier design phase, but it is more usual for these to follow on after the contract has been awarded. It is common to wait until this point in case there are changes made to the detail or to the specification at the contract award stage. The provision of detailed working information is one of the major tasks in any building project, and is one which can often lead to mistakes, delays and consequent arguments. Information on detail is very much within the scope of work as far as the project manager is concerned; it has significant cost in itself, it uses resources, takes a long time, and the quality of information must be controlled. Most professionals in the construction industry will be well aware of the amount of effort that has to be put into the preparation of working details and also the time that is spent 'chasing' it and arguing about it; there is a real challenge here to project managers to solve this problem area.

The construction phase   This phase of a project is often seen as the main focus of attention, probably because it is the most visible, but also because it costs the most money. It does not however involve the most important decisions which are almost certainly taken at the stages of choosing a designer, briefing, design, project appraisal and contract award. The construction phase may be the longest, but not necessarily so, and in most cases only represents a minority of the whole project duration. It does involve a very large number of interdependent tasks, and hence needs to be planned in great detail. Much attention is given to the planning of site work, which is discussed in more detail in several other chapters.

Project close-down   Close-down is the phrase that is often used to describe the tying up of all the loose ends of a project, and is one area that is often not given enough attention. In a construction project it will include such things as a final measurement and final account for the works completed. There will be 'as built' drawings to be prepared and submitted to the client for subsequent use in maintenance work and modifications. It is also necessary to check that all work has been completed satisfactorily, that is to say the project management has been thorough. In many cases this may be the end of the construction contract but it is not the end of the project; there may be many other activities concerned with the commissioning, occupation and use of the building, the recruitment of staff, and the whole process of getting the plant into production. If all of this is to be included in the project it is probable that the project management will be carried out by a representative of the client or user company; for such a project manager all these later stages of work will be within the scope of the project.

Ultimate disposal   Although rarely considered to be part of the project, ultimate disposal of the constructed works is in a number of special cases relevant. In recent times two important examples of this are nuclear power stations and off-shore oil structures, both of which involve expensive and difficult de-commissioning. Where the possibility of such work is seen it should be considered to be part of the project, especially in regard to the initial project appraisal where long-term liabilities should be taken into account.

The project scope   The scope of a project may include some or all of the phases listed above, and it is important when setting down the responsibilities of the project manager that these should be made quite clear; in other words the scope must be clearly defined. One thing which is important to note is that most of these stages will exist, irrespective of whether they are placed within the scope of the responsibility of the project manager or not. If they are omitted they will have to be effectively managed by someone else, or there is a danger that they will not be properly carried out. For example if the initial stages of concept and briefing are allowed to drift on, there will be an inevitable delay in the project completion. There is a real danger that formal project management will only be applied to the main construction phase since this is the period where most major expenditure occurs. The pre-contract work may well occupy at least as much time, with important consequences for programme, and similarly the occupation and operation of the built facility will also need a long period. It is essential in projects of the type described in this chapter that clients and their consultants fully appreciate that each of the stages listed above and shown in Fig. 5 do take significant time to complete, and that a delay in any one of them will lead to a consequent delay in the project. Clients in particular are very prone to taking the view that early decisions can be delayed or changed simply because no work has started on site and 'there is plenty of time to catch up'. This is a total myth; once time has been lost it cannot be recovered. So-called catching up can only be done at the expense of something else, either money, quality, or time for other tasks. It cannot be emphasised too strongly that all phases in a project should be properly planned and managed. This realisation may come as a bit of a surprise to clients who feel that since they are paying for the project they have the right to change their minds, add to or delete from the scope, take as long as they wish over decisions, and still expect the project to be completed on time. They may certainly have some of these rights, provided that the contract allows for them, and provided they fully appreciate the impact upon cost and completion date. It is useful to remind clients, designers and contractors alike that in terms of time the client will primarily be interested in the time taken to get from the initial concept of the project right through to the point where the facility is in full operation. The relative lengths of pre-contract, contract and post-contract periods are of little importance except insofar as they affect the cash flow timing.

The real message of the above paragraph is that every step in a project must be completed before its successor can begin, every step will take time and will need resources, and that delays at the early stages of a project will have a knock-on delaying effect. It may be difficult for the project manager to convince the client of this fact, but it must be faced. For client companies the important point to note is that effective project management should be put in place as early as possible, certainly by the time that the initial gleam of an idea has been put down as a proposal for consideration. Delaying the appointment of a project manager until the proposed development has received company approval is already a little late, as the project clock will have been ticking for some time.

'Project life cycle' is a phrase which covers the whole period from initial concept through to steady-state use. In the case of construction this may be extended to include the whole life of the facility including subsequent modification and ultimate dismantling and disposal, although these aspects might be better thought of as being later and separate projects. In other projects not involving construction ultimate disposal is seldom a problem. The similar phrase of 'product life cycle' in manufacturing has a rather different meaning; it covers the initial idea for a product, its development and design, setting up for production, its useful life as a saleable product (*not* the life of an individual item produced), the decline in demand with obsolescence, and finally termination of manufacture. It is of interest to note that the early stages of a product life cycle, up to its market launch will have the form of a project.

A case of a new consumer product

Photo Products Ltd, a company which has been in the business of making materials for the photographic market, has carried out research and development work which has given rise to a totally new process which is not in use elsewhere. The process, however, only works with a new design of camera, and therefore the market will be restricted to a small number of users, and might easily be suppressed by existing competitors. The company has come up with the idea of entering the market for cameras, even though it has no existing facilities or expertise in equipment manufacture. Development work on the camera has been undertaken as part of the work on the process, and there exists a development prototype which has produced good results. The benefit of entering the equipment market lies in the ability to sell a photographic system as a whole, with no identical competition, based on features which are unique to the system. The implementation of the proposal can be regarded as a project with the following stages to be included. This is not an exhaustive list and the project is expanded as an exercise later in various parts of this book under the title Photo Products Ltd.

The detailed proposal phase for this project   The proposal phase would include market research, design studies, manufacturing studies, review of patents and licences, and estimates of capital and operating costs. At the end of this stage it would be usual to make a firm proposal to continue with the project into the detailed development, possibly only as far as the point where capital investment on land and equipment would be required.

Detailed development   Development would then proceed and would involve more commitment to expenditure of time and money on such tasks as

- the working design of the product and the production methods
- determination of materials and components suppliers
- marketing and distribution arrangements
- quality management systems
- administrative procedures including data processing
- staffing requirements
- seeking a suitable site and identifying appropriate plant and equipment
- preparation of capital and operating budgets
- determination of sources of finance
- assessment of necessary consents.

When all this is complete and reported upon, a final decision would be required before the implementation phase commenced with consequent major expenditure.

Implementation   This will include site acquisition, and construction of the building; the ordering, delivery and installation of plant and machinery for both camera and processing materials. Packaging design and preparation, marketing agreements, storage and distribution, staff recruitment and training, production test runs, test marketing, trade publicity and many other tasks would follow and complete the main part of the implementation project with the public launch of the new photographic system.

The close-out   This is the recognised project management jargon for the final stage of the project with a review of how performance had compared with plan, and what lessons could be learned for the future. This stage should also include documentation of the whole project so that operating managers have full information on the plant and process which will be needed for maintenance, modification and/or renewal.

A case study of a new information processing system

In an existing company it is felt that the present information system is inadequate, and that the simple replacement of present equipment will not meet the needs of the company. Often the starting point for a project of this type will be an in-house temporary project team of staff members who are asked to look into the idea. It is possible that this team will have the necessary skills, but it is unlikely that they will have experience of a range of appropriate systems elsewhere. What might well emerge is that the project team recommend the appointment of a consultant to advise on how the company should proceed. If this is done, the first task of the review team and the consultant would be to prepare the brief for the project, including a definition of the responsibility for the project management, to form part of the definition of the project scope. The detailed stages of the project would then include the following range of activities.

Determination of information needs   Problems with existing systems, design of a new system including both hardware and software, review of similar existing systems elsewhere, estimates of capital and operating costs, and other items will all be required before a decision to proceed can be taken.

The detailed design   The design of procedures and documentation, the selection, purchase and installation of hardware (often including some building work), the writing and testing of software, the training of operating staff and then the final installation and commissioning of the new system will all be included in this stage. There will normally be a close-out of the project which records the process and decisions of the project and provides full information on the new system to help in its subsequent use and maintenance. As in the two earlier projects described, the sum of all these tasks will constitute the scope of the project which must be clearly set out with responsibilities defined.

Information projects have a number of special characteristics which may not apply to construction and manufacturing plant. One of these is that it may take some time for the client to fully understand and use the capabilities of the new system, and it is important that the scope should include continued support from specialists over a long period. The same is not true of a building project where the user can easily see what the building includes and what use can be made of it. Another dilemma is whether initially to seek the advice of a consultant who has knowledge of a range of makes of hardware, or whether to select a manufacturer on the basis of reputation and experience, or even to seek competitive bids from a number of suppliers. The difficulty of the last is to know how to compare bids, as they will almost certainly be based on different technical performance. Perhaps the best approach is to get expert advice on the preparation of a brief, then seek bids which meet that brief and definition of scope, and select the cheapest bid (counting both purchase and operating costs) which fulfils the scope in content and performance. There are several good books on the project management of information systems (Bentley, 1982; Lewin, 1988; Rakos, 1990).

5 The management of time

This chapter gives a step-by-step description of the calculations involved in networks. Those readers who are familiar with either arrow or precedence diagrams can omit the detailed sections or use them for quick revision. For those who are totally new to networks it is recommended that they take the time to work through these sections and also several worked examples. Understanding network plans, even if these are produced by others, does depend on a good basic knowledge of the principles and the terminology used. This chapter describes both arrow diagrams and precedence diagrams, but most readers will only need to learn one or the other. Each is therefore considered separately, which means that there is inevitably some repetition in the text; it is hoped that readers who do look at both methods will not find this repetition to be irksome. For those who wish to concentrate on arrow diagrams the description of these comes first, followed by virtually the same examples in terms of precedence diagrams. If readers are already familiar with one of these two methods it may be valuable to read the earlier brief section on linked bar charts, since this is now a common form of presentation, especially when a computer package is used.

In Chapter 3 the management of progress was discussed in terms of time, cost, and more importantly in terms of real progress. The purpose of this chapter is to look at the management of what has been completed in terms of the working time taken to do it (measured in worker-hours), and the time elapsed in completing a project. It is important to remember that it is essential to distinguish between these two aspects of time. The fact that project work comprises a large number of different and interdependent items of work makes the counting of items inapplicable as a means of planning or measuring work. The procedure developed to measure progress is to create a diagrammatic model in which each item of work is represented by a symbol, and to link these symbols in a way which clearly shows the sequence in which the work can be done. This is the technique which is known as network analysis or network planning. Strictly speaking the analysis must precede the planning as explained in Chapter 3. At one time networks were thought to be all that was needed to manage a project, and while this is now known to be wrong it does not diminish the value of networks. Simply, it is now realised that there are many other techniques which must also be used in parallel.

ARROW DIAGRAMS AND PRECEDENCE DIAGRAMS

There are basically two quite distinct methods of drawing networks, namely 'arrow diagrams' or 'precedence diagrams'. Both methods are perfectly valid, but it is important not to confuse the two; like driving on the left or right side of the road, a concensus must be reached. In order to avoid confusion here, the first description will cover arrow diagrams, and then in a later separate section the calculations will be repeated for the case of precedence diagrams. It is suggested that for most people it will only be necessary to become familiar with one of the methods, and indeed it may be helpful only to learn one, at least at the outset.

Once you have work experience of one method it will be easy to understand the other. Many articles have been published that extoll the virtues of each of the two methods and explain why it is preferable to the other; in truth there are advantages both ways and it is a matter of the preference of the organisation which is preparing the network. For any individual it is perhaps best to try to find out which of the two methods is being used in your organisation and/or by the other companies with which you are likely to be working. An examination of a range of project management books shows that most of them use arrow diagrams, and relatively few use precedence diagrams, and for this reason most of the examples in this book are based on the arrow method. The use of precedence diagrams became more popular with many of the computer packages, but as these have developed and become more user-friendly the current thinking in many applications is to use a third slightly different format, namely that of a linked bar chart which is effectively a way of bringing the two together. This is also described later in the text and can easily be developed from either arrow or precedence diagrams. However, it is probably best to leave this until a later stage and to get to grips with network logic either through arrow or precedence diagrams. What is vital to learn is the development of sound basic logic, without building in preconceived ideas. Many years of teaching project management to a wide range of students and professionals has brought a realisation that many people will too readily jump to conclusions about the way a project should proceed. The following section may appear a little pedantic at times, but all too often it is found that the management of projects runs into difficulty not because the methods are wrong but because they have been wrongly used. In order to avoid the trap of blaming our tools let us consider firstly the way in which an arrow-based diagram is prepared. The next few pages follow closely the author's earlier text which has been found to be useful (Woodward, 1975). Another way of self-teaching is by working through a programmed text (Silverman, 1988).

Arrow diagrams

A project comprises a large number of separate items of work which we refer to as either jobs or activities and which we can regard as not requiring further sub-division. We then represent each activity by an arrow which says only a limited number of things about it, as shown in Fig. 6. This states that at the tail of the arrow work can start (but does not necessarily do so), and by the head of the arrow work must be

Fig. 6. *What an arrow represents*

complete. The arrow is not normally drawn to a time scale, and it is not implied that work will be continuous over the length of the arrow. This has the advantage that it is not necessary to consider the time aspects of the project at the outset, and it is more important to get the logical sequence correct in the first place. This avoids the temptation of writing into the network what you would like to happen rather than what really must happen. There is no particular significance to the actual orientation of the arrow. It is usual for the general progress of the project to run from left to right, although some arrows may run the other way or cross over the lines of others. Remembering that we are building a model to represent the project, a reasonable way is to begin at the beginning, i.e. to consider which activity can be carried out first and then draw down an arrow to denote it, writing a brief work description along the arrow. The next step is to ask whether there is any other activity which must be completed before the one we have drawn, and if the answer is 'yes' it should be shown in the diagram immediately prior to the first one drawn. If the answer is 'no' we can be sure that we have correctly identified the start of the project. There are three questions that should be asked, including the one we have already dealt with.

(a) What other activities must be complete before this one can start?
(b) What other activities can be done at the same time?
(c) What other activities cannot begin until the one we are looking at is complete?

If any activities can be carried out at the same time these should be shown in parallel, as in Fig. 7, and we indicate all those which can start as soon as we have finished the first one, by drawing them immediately after the first arrow. The whole of the network proceeds on this way, remembering that we are setting down the essential logic, not a sequence that we would 'normally' follow, or one that we would simply like to follow. It is important that the model is truly a representation of what must be done, or we will lose much of the value of the method.

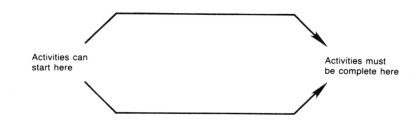

Fig. 7. *Activities which can be carried out at the same time*

In order to see the importance of this rigour consider a simple example of the erection of a steel portal frame (Fig. 8) in which the following activities have been identified.

(a) clear site
(b) prepare base A
(c) prepare base B
(d) erect column A
(e) erect column B
(f) erect the beam
(g) check the alignment
(h) fix the frame

It is fairly obvious that (a) is the first activity, but this can be confirmed by the first of the three standard questions. The second question will confirm that nothing else can be done at the same time, and the third question will tell us that the preparation of bases A and B can both proceed in parallel as soon as the site has been cleared. The same rigorous questioning will lead to the completion of the network as shown in Fig. 9, in which we note that the erection of column A depends only on the preparation of base A and not on base B. The erection of the beam does require both columns to be in position, and this is shown by the two arrows coming together at that point. The other activities follow one after the other, checking the logical dependencies at each step. The network should show what is reasonably possible without taking special measures at any stage. For example, it would technically be possible to erect the beam without column B being in place by using temporary supports, but this would mean using an expensive and unusual procedure which would only be adopted if there were particular requirements to do so. This procedure may appear to be fairly straightforward, but it does not always turn out that way in practice, as can be seen from the following.

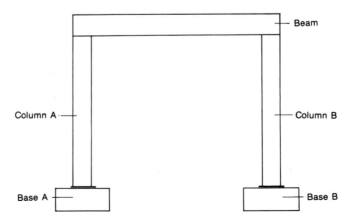

Fig. 8. Erection of portal frame

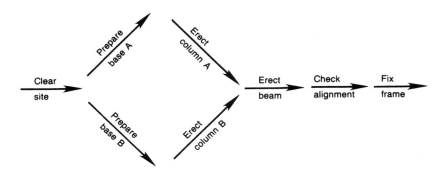

Fig. 9. General arrow network for portal frame project

There is a strong likelihood that each planner of a significant project will prepare a network slightly different from one prepared by colleagues, and this can be seen even in the trivial example above. In the portal frame case it could be argued that there would only be one gang of workers available to prepare the two bases, with the implication that base B would follow on after base A rather than run in parallel, and would appear on the network as shown in Fig. 10. This would imply however that it was essential for A to precede B whereas they could be in the reverse sequence. Putting the two base preparation activities in sequence is a matter of convenience rather than a technical requirement and should not be built into the network. At the same time it could be argued that on a real site it is likely that the two bases would firstly be excavated by the same machine on the same day, followed by a ready-mix truck filling them both with concrete at the same time. This could be shown by a different version of the network, as in Fig. 11, but this would imply that it was

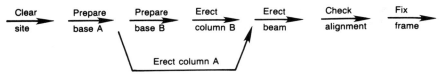

Fig. 10. Arrow network — portal frame project, bases in sequence

Fig. 11. Arrow network — portal frame project, bases at the same time

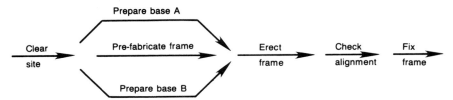

Fig. 12. Arrow network — portal frame project with frame pre-fabricated

essential for technical reasons that both bases were constructed at the same time; this is unlikely to be true. Further versions of the network could be drawn to show the two columns following one after the other, or yet another to show the frame of two columns and a beam being pre-fabricated and then erected as a whole unit, as the network in Fig. 12. Each of these later network diagrams represents one particular approach to the construction and is therefore a specific plan rather than an analysis of the problem. This is an example of jumping to a design without carrying out any calculation and should be avoided. The correct logic is that shown in Fig. 9, namely showing the essential, and only the essential, sequences.

Another basic point in the preparation of a network is the level of detail to which it should be taken, i.e. how big is an activity? In the above example it could well have been the case that the whole of the erection procedure would be covered by a single activity, or could even be part of a larger one such as 'build pipe bridge' which might easily comprise 20 portal frames. There is no simple way of knowing the correct level of detail, and it can only be said that it should be appropriate to the plan in hand and that experience is the best guide. One very rough guide to deciding the level of detail is to say that no activity should have a duration of less than one hundredth of the project duration, but there will always be exceptions to this. What is likely to happen in practice is that the master plan for a large project would have a single activity 'build pipe bridge', the sub-contractor building the bridge and fixing the pipes might have a separate activity for each portal, but the actual work gang would need to have a more detailed plan for a least one typical portal.

It can be seen that each network should initially be drawn on the basis of these simple guidelines.

(a) Decide on the level of detail at which activities will be defined.
(b) At the initial stages do not consider time at all, only the logical sequence of activities.
(c) Start every activity as soon as possible and finish it as late as possible.
(d) Only make activities follow those which are essential, not simply those which are convenient or desirable.
(e) In order to avoid prejudice make an initial assumption of unlimited resources — this can be corrected later, but will avoid false logic in the original diagram.

It has been shown above that each activity is indicated by an arrow. In a diagram of this form the point at which two or more lines meet is referred to as a 'node', and in the case of an arrow network this represents an 'event', i.e. an intermediate stage in a project where some activities are complete but others not yet started. It is normal practice to give each of these events a unique identifying label, usually in the form of a number. For example, in Fig. 9, event 5 is the stage in the project where both column A and column B have been erected, but no work has yet started on the erection of the beam. By using this numbering system it is possible to identify each activity by a pair of numbers, i.e. the event number at its tail, followed by the event number at its head. This would mean that 'erect beam' would be indicated by the pair of numbers 5 and 6, and would be called 'activity 5–6'. It is of interest to note that if a full list of activities is given in this way it is possible to reconstruct the logical sequence simply from the event number pairs; in fact this is the way in which a computer would understand the logic of a network. For this reason it is usually required that there is no duplication of event numbers, and it is a good idea that for each arrow the two event numbers are in ascending order (but not necessarily in sequence). It is worth trying the very simple exercise of reconstructing the logic of Fig. 9 from the list of activities in the project.

- 1–2 clear site
- 2–3 prepare base A
- 2–4 prepare base B
- 3–5 erect column A
- 4–5 erect column B
- 5–6 erect Beam
- 6–7 check the alignment
- 7–8 fix the frame.

This may all sound a little tedious at this stage, and several computer packages are able to dispense with such rigour; however the discipline of working in this way will help generate the right basic approach and give a good grounding in the use of networks. Experience of teaching the drawing and use of networks has shown that it is not always easy to grasp all this at the outset, but once understood it is easily remembered — analogous to riding a bicycle!

Dummy activities

These are used in arrow diagrams in two separate situations. (Note that we rarely talk about dummy jobs, but in all other uses the words activity and job are synonymous.) First in the allocation of event numbers there will be situations in which two separate activities both start and finish at the same event, e.g. in Fig. 13 the purchase of software and the installation of the computer in the office could proceed at the same time, and both are required before the machine can be operated. As shown however both of these activities would have the event number pair '2–3', and could consequently be confused. The way to overcome this is to introduce a dummy, shown as activity 3–4 in Fig. 14. The two activities are now uniquely identified as

(*a*) 2–3 buy software
(*b*) 2–4 install computer in office.

Note that the dummy is shown as a dotted line, and it does not take any time to complete. It simply transfers the logical sequence forward and in subsequent calculations is regarded in the same way as any other activity.

The second and more important use of dummies is where they form an essential role in maintaining strict logic. Consider the following logic which forms part of another project

- order and install fuel tank
- order fuel
- paint tank
- take delivery of fuel.

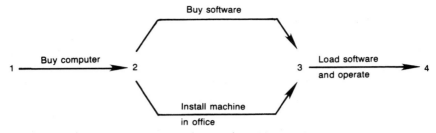

Fig. 13. Illustration of non-unique numbering of activities

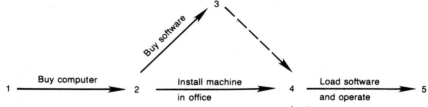

Fig. 14. Use of dummy activity to ensure unique activity numbering

A first attempt at the logic diagram for this might look like Fig. 15. This would correctly show that the tank could not be painted, nor the fuel delivered, until the tank is installed. However, it would also state that the tank could not be painted until the fuel had been ordered, and this is clearly not essential. This problem is overcome by the use of a dummy as shown in Fig. 16 which now indicates that both the tank installation and the fuel order are needed before the delivery of the fuel, but that only the tank installation is necessary before the tank can be painted. This procedure is important because it is wrong to build in false dependencies which would mean that

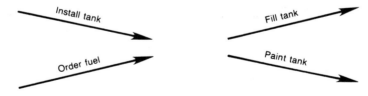

Fig. 15. Non-essential dependency has been shown in logic

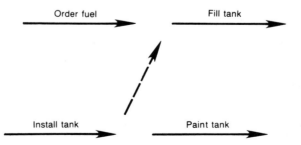

Fig. 16. Use of dummy activity to avoid false logic

the network model no longer represented the reality of the project. A more realistic example is illustrated in Fig. 17, which shows the network for part of a building project; this is correct as it stands in that all four activities leading to the central node, i.e. event 22, are necessary for the two following activities of installation of power cables and finishes to walls and floors. This is the stage of the building where it is effectively watertight, which is essential for the two following activities if damage is to be avoided. It is now intended to add a third activity, that of building an interior partition wall, starting from the same event. If this wall is to be of a material which does not need to be protected from the weather there would be no need to wait for the building to be weather-tight, and the partition could be started as soon as the exterior walling is in place. It would be wrong to insert the arrow for interior walls starting at event 22 since this would imply that all other activities ending there are essential for these walls. The solution is to insert a dummy arrow 21–22, as in Fig. 18, which then states that building the interior walls only depends upon the completion of the exterior sheeting. (There would be other requirements which are not shown in this diagram, such as the construction of the floors, availability of materials, detailed information and so on.) It is always worth looking closely where a whole cluster of jobs merge into an event and/or emerge from an event; what this states is that every activity starting there requires the completion of all activities finishing there, a situation which is not always a true representation of reality.

Drawing an arrow diagram

The drawing of an arrow diagram requires a disciplined approach. Each project planner will develop an individual format or style to suit the needs of the project

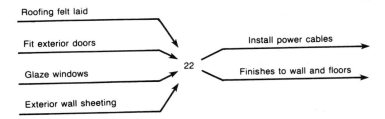

Fig. 17. Initial logic, before adding extra activities

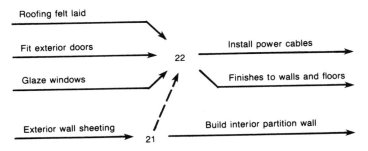

Fig. 18. Use of dummy activity to maintain correct logic

and/or the organisations involved; the following suggestions may be useful in guiding this.

- There is little point in writing out a long list of all the activities, as the network can effectively be the list. Once the planner has a little experience there is no need to prepare a rough network with the intention of having it redrawn; indeed the introduction of another step may simply provide an opportunity for errors to creep in. After a little practice it should be possible to prepare a reasonable network at the first attempt. In many situations the network may now be effectively 'drawn' directly into the computer, indicating the logic relationships as it proceeds.
- At the start the planner should work in terms of logic only, leaving durations to be entered later.
- The network, if drawn by hand, should be in pencil and at a reasonable scale, since it is almost certain that alterations and additions will have to be made as the work proceeds. There have always been jokes about networks only being used to paper the walls of the office, but do not be deterred by such jibes — it is likely that once finished most of the subsequent work will be done on a computer, and the large drawing will only be looked at on a few occasions.
- It is useful on the drawing to show each arrow with a horizontal somewhere along its length, with the activity description written out along that horizontal; this makes the whole thing not only tidier and easier to read, but also permits 'strings' of related work to be in one line, e.g. for an area or for one type of work. The procedure of using a series of code letters for activities cross-referred to an index should be avoided, as it makes reading the network very tedious and can lead to errors.
- Remember that the network should be drawn initially on the basis of what is possible rather than what is intended, it should make no assumptions about resource limitations and should not be based on one specific method of implementation.

Estimating durations and calculating project time

Up to now the network has been drawn without regard to time in order to ensure that the logic is not distorted. It is now necessary to fix durations, and as a first approach an estimate is made for each activity. For convenience the time unit to be used in each project is selected such that it is possible to work in whole numbers of that unit. For example, it is common to use either a week or a month in the case of an outline network for a major project, commonly a day is an appropriate unit for a working network for a building or a computer project, and perhaps an hour for a machine overhaul. It is generally recognised that estimates can only be approximate because of the absence of a direct precedent in the case of project work, and for the immediate discussion a single time estimate is used. There are ways of dealing with variability in activity durations, and this is considered in Chapter 13. The estimate of a duration should be made wherever possible on the basis of prior experience, taking account of any variation in conditions. Durations are expressed in terms of elapsed time, as was discussed earlier in this chapter, but some care has to be taken with this. Most activities will only progress during working hours whereas others simply represent the passage of time, e.g. a computer programmer will only work on a project for say 50 hours per week, while the delivery of a new piece of hardware might be expressed in terms of weeks. It is necessary to express both of these activities in the same units, not simply in hours but more specifically working hours. In order to do this a decision must be made on the number of working hours in a week, such as 50 in the case just quoted, and all durations then quoted in that unit; e.g. a delivery of 2 weeks would become 100 hours.

The procedure for calculating project duration is shown in Fig. 19. The particular form of notation used in this text is commonly but not universally in use. In many cases computers will be used to carry out calculations, but this is not essential, and it is useful to know the notation which is generally used in networks calculated by hand. On the arrow network diagrams in this book the symbols generally used are described below.

The 'duration' of an activity is shown below or beside the arrow near its mid-point and is enclosed within round brackets, e.g. (24) would mean a duration of 24 time units. The 'event number' is simply written in the space between arrow heads and tails at the node. The 'earliest event time' of an event is the earliest time that the event can take place, and is shown in a square box, e.g. $\boxed{14}$. Sometimes in description this is referred to as 'EET'. The 'latest event time' is the latest time at which an event can be allowed to occur without delaying the project, and is denoted by a number in a circle, e.g. $\textcircled{8}$. This is often referred to as 'LET'.

The calculation of the project duration The project starts at time zero and hence we put this as the earliest event time (EET) for event 1. Activity 1–2 has a duration of 2 days, assuming we have selected a day as the time unit, and therefore earliest time that event 2 can occur is $\boxed{0}$ + (2) = 2, and this is then entered in a box at event 2. Proceeding through the network, we can now calculate the EET for event 3 is given by $\boxed{2}$ + (3) = $\boxed{5}$, and so on. It is necessary in calculating the EET for any activity that the EET for all preceding activities has been entered into the diagram.

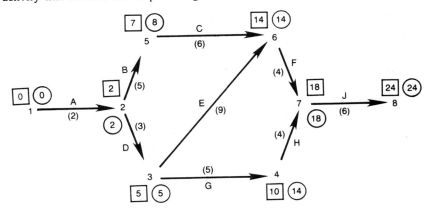

Fig. 19. Network showing calculation of event times

Care has to be taken with events where two or more activities converge, e.g. event 6. Here we have to consider two activities, 3–6 and 5–6, since both have to be finished for event 6 to occur. The EET for event 6 is given therefore by taking the later of the EET at event 5 plus the duration of activity 5–6, or the EET of event 3 plus the duration of activity 3–6, thus

$$\boxed{7} + (6) = \boxed{13}$$
$$\boxed{5} + (9) = \boxed{14}$$

We must choose 14, the later of the two, since we must allow enough time for all activities leading to event 6 to occur. This will happen at all converging events, the only other one in this network being event 7. We are thus able to conclude that the EET for the finish event 8 is 24 days, and this gives us the minimum project duration.

We now carry out a reverse calculation to establish the latest time at which each event can be allowed to take place, without delaying the project, and this is done by starting at the finish event of the project, i.e. event 8. If we are to be at event 8 at time 24, event 7 must be reached 6 days earlier since that is the time needed to carry out activity 7–8. The calculation of latest event time (LET) at event 7 is given by $\textcircled{24}$ – (6) = $\textcircled{18}$. We continue through the network in reverse order of event number, but again have to be careful where two activities emerge from an event, e.g. event 2. There are two paths to be considered here, and both must be calculated

$$\textcircled{8} - (5) = \textcircled{3}$$
$$\textcircled{5} - (3) = \textcircled{2}$$

In deciding which of these two to use we should remember that enough time must be given to complete all activities, and we must therefore take the earlier of the two alternatives, namely day 2. Similar steps would have to be taken at event 3. It may at first seem strange when calculating the LET to select the earliest alternative, but this can be understood by considering the two sequences of activities which have to be completed after that event. One path is through events 2, 5, 6, 7, 8. with a total duration of $(5) + (6) + (4) + (6) = 21$, and the other path is through events 2, 3, 4, 7, 8, with a total duration of $(3) + (9) + (4) + (6) = 22$. There must be enough time for both of these paths to be completed, and this will be determined by the longer of the two, i.e. 22 days. The process of calculating the LET by working backwards through the network is completed when the start event 1 is reached, at which the LET should be the same as the EET, namely zero. If this is not the case it means that an error of calculation has been made, and this can therefore serve as a useful, but not total, check on the calculation. Remember that when calculating the latest event times of any event it is necessary to have already worked out and entered the LET for all following activities.

The critical path in an arrow diagram   It is worth reflecting that the total project duration is determined by the sequence of activities which requires the longest time, and it is therefore the longest path through the network, not the shortest. Comparison of EET and LET at all events will show that there are several events at which the two values are the same, and furthermore they form a continuous path through the network. This means that at each of the events there is no time to spare since the earliest possible time coincides with the latest allowable time, and the project is critically dependent upon these events occurring at that time. The path through these critical events is called the 'critical path', a phrase which applied initially to the whole subject area, and became known as the 'critical path method', usually shortened to CPM. In more recent times it has been realised that while the critical path still features strongly in the management of projects it no longer covers the whole subject area which is now known as project management.

The critical path must not be brushed aside, since it gives us a very useful means of focusing attention in a project.

- Although it is the longest path it gives the shortest possible duration of the project. In practice it is seldom possible to complete a project in this time because of limitations on the availability of resources, unforeseen delays and so on (see Chapter 9).
- All critical activities must be completed on time if the project is to be finished on time. If project completion is to be accelerated then the critical path must be shortened; note however that reducing the length of the critical path is not in itself a guarantee of early project completion (see Chapter 8 on project acceleration).
- The critical path does serve to concentrate the minds of those who are responsible for activities on that path, ranging from the client making a decision at the start, through the designers and construction team, and finally to the finishing and hand-over.
- All activities on the critical path have zero float (see below for definition of 'float').

Float in arrow diagrams

The preceding section concentrated on critical activities, i.e. those with no time to spare. There are however others in a network which are not critical and do have more than enough time for their completion, e.g. activity 3–4 in Fig. 19. This spare time is referred to as 'float' which is a useful concept, and can be explained as follows. It is the difference between the time available for an activity and the time needed to complete it. There are two main types of float which are commonly used, 'total float' and 'free float' and these are calculated in the following way.

The total float of an activity is given by the LET at its finish, minus the EET at its start, minus its duration. In the case of activity 3–4 in Fig. 19 this would be ⑭ − ⑤ − (5) = 4. What we have done here is to take the maximum possible available time by assuming the earliest possible start and the latest possible finish for the activity, and then compared this with the required duration.

The free float on an activity is defined slightly differently, and actually forms part of the total float. Free float is given by the EET at the end of the activity (not the LET as in the case of total float), minus the EET at its start, minus its duration. In this case we have not taken the maximum available time but the time available without causing any delay to a subsequent activity. Referring again to Fig. 19 this would be the calculation for activity 4–7: ⑱ − ⑩ − (4) = 4. As it happens the total float for

71

activity 4–7 is also 4 days, but this is not a coincidence; on a non-critical string of activities where it rejoins the critical path the total float on activities in that string will appear as free float on the last activity in the string.

All this may appear to many readers as an over-indulgence in the manipulation of numbers, but with a little practice it is not difficult to become familiar with the concepts of total and free float and the difference between them. The good news is that it is rarely necessary to carry out the calculations by hand, since this is easily done by computer; it is important though to know what they are and how they can be used. The management of time in projects really comes down to the manipulation of durations and floats and is an essential part of project management. Several examples of this will be discussed in later sections and chapters, but before reading these it is advisable to become familiar with the terms and calculations by spending some time on the exercises given in the appendices.

Activity dates

It is now necessary to introduce the concept of activity dates, as distinct from event times. An event time is a point in the network where some activities have been completed, but others not yet started. For example, in Fig. 19 the EET of activity 2–5 is shown as day 7, and this really means 'the end of day 7', a specific point in time which actually coincides with the beginning of day 8. This does at first seem to be confusing, but can be explained by the fact that each numbered day actually has a duration of one day, and the beginning and end of any one day are in fact a day apart. We now look at activity start dates.

The 'earliest start date' (ESD) of an activity is the earliest date on which work can commence on that activity; in the case of activity 5–6 in Fig. 19 this would be day 8, i.e. work could begin on the morning after day 7 which is the EET of event 5. The 'earliest finish date' (EFD) of an activity is the earliest date on which work on that activity can be complete; for activity 5–6 in the same small network being considered, that would be day 13, which is 6 days later than the EET at event 5. Note that it is not necessary in the case of finish dates to move on to the beginning of the next day; this is because if a finish date is specified as being day 13, then it would be interpreted to mean the end of day 13. The 'latest start date' (LSD) of an activity is the latest date on which work can start on that activity without causing a delay to the project. For the same activity 5–6 this would be day 9, which is the day after the time given by deducting the activity duration of 6 days from 14 which is the LET of event 6. The 'latest finish date' (LFD) is simply the same as the EET at the end of activity 5–6, namely day 14.

All this complicated arithmetic can be more easily understood by looking at Fig. 19. It is of importance to know the meaning of start and finish dates since these are usually the figures that are printed out by a computer. It should also be noted that the start and finish dates have the same meaning and values irrespective of whether they are calculated from arrow or precedence diagrams for the identical project.

Familiarity with diagrams

This ends the section dealing specifically with arrow diagrams, and marks the start of the description of precedence networks which is largely a parallel of the section just completed. It may be a good idea for readers not familiar with networks to concentrate initially on arrow diagrams and carry out practice calculations until these are familiar. There is a real possibility of confusion if the above section on arrow diagrams and the following one on precedence diagrams are studied together before one of them is fully understood.

Precedence diagrams

As stated earlier, precedence diagrams form an alternative to arrow diagrams, but the two should not be confused. In order to avoid confusion in this book the following description partly repeats what was said earlier, so that there is no need to refer back to the previous section. Any readers who do not wish to follow this next section on precedence diagrams can omit it; they should resume reading from the section headed 'Other forms of float' (page 81).

This first section dealing with precedence diagrams is based on the same logic principles already outlined for arrow diagrams, i.e. it is assumed that each activity must be totally completed before its successor commences. There are situations however where it may be possible and desirable to commence the second activity some time after the first has started, e.g. in the case of building a long brick wall it would be intended to commence brickwork a few days after work on the foundation had started, once sufficient length was finished. This situation can be handled using either arrow or precedence diagrams, and it is discussed more fully at the end of this chapter.

Precedence diagrams look rather like logic diagrams as used in computer or systems analysis. In this case the project is divided into a number of activities which are not further sub-divided, and each one is represented by a node or box in which certain information is noted. It is for this reason that precedence networks are sometimes referred to as 'activity-on-node diagrams'. For the purposes of illustration we shall consider the portal frame project of Fig. 8, in which the list of activities was

(*a*) clear site
(*b*) prepare base A
(*c*) prepare base B
(*d*) erect column A
(*e*) erect column B
(*f*) erect the beam
(*g*) check the alignment
(*h*) fix the frame.

Each of these activities is now shown in a box as in Fig. 20 and it is necessary to show the logical sequence connections between them. This can be done by asking two questions about each activity.

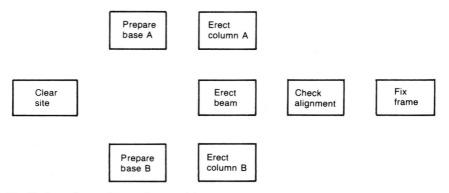

Fig. 20. Precedence diagram for portal frame — activities only

(a) What other activities must be complete before this one can start?

(b) What other activities cannot start until this one is complete?

The sequencing is then marked on to the network drawing by inserting linking arrows between boxes. It is fairly obvious that 'clear site' will precede all other activities and therefore no arrow is shown leading into that box. Once the site is cleared it is possible to start work on the preparation of the two bases, and this is indicated by two sequencing arrows, one from 'clear site' to 'prepare base A', and the other from 'clear site' to 'prepare base B', as in Fig. 21. The questions are then asked about each activity in turn, starting from one that has already been linked into its predecessors. 'Erect column A' will clearly follow 'prepare base A' and so on through all the other activities. The network should show what is reasonably possible without taking special measures at any stage. For example it would technically be possible to erect the beam without column B being in place by using temporary supports, but this would mean using an expensive and unusual procedure which would only be adopted if there were particular requirements to do so. This procedure may appear to be fairly straightforward but it does not always turn out that way in practice, as illustrated in the following paragraph.

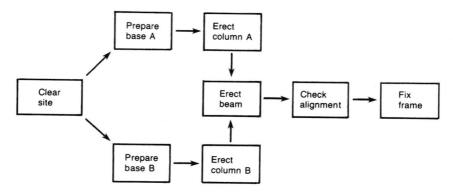

Fig. 21. Precedence diagram for portal frame — activities and sequencing arrows

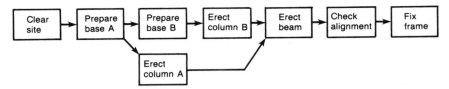

Fig. 22. Precedence diagram for portal frame — bases prepared in sequence

There is a strong likelihood that each planner of a significant project will prepare a network slightly different from ones prepared by colleagues, and this can be seen even in the portal frame project. In this case it could be argued that there would only be one gang of workers available to prepare the two bases, with the implication that base B would follow on after base A rather than run in parallel, and would appear on the network as shown in Fig. 22. This would imply, however, that it was essential for A to precede B whereas it could be in the reverse sequence. Putting the two base-preparation activities in sequence is a matter of convenience rather than a technical requirement, and should not be built into the network. At the same time it could be argued that on a real site it is likely that the two bases would be excavated by the same machine on the same day, followed by a ready-mix truck filling them both with concrete at the same time. This could be shown by a different version of the network, as in Fig. 23, by making a single activity of the two bases, but this would imply that

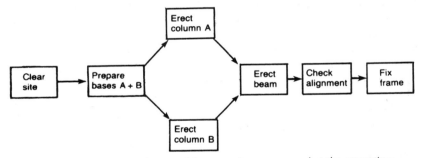

Fig. 23. Precedence diagram for portal frame — bases prepared at the same time

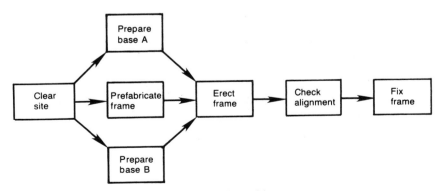

Fig. 24. Precedence diagram for pre-fabricated portal frame

it was essential for technical reasons that both bases were constructed at the same time; this is unlikely to be true. Further versions of the network could be drawn to show the two columns following one after the other, or yet another to show the frame of two columns and a beam being pre-fabricated and then erected as a whole unit, as in the network of Fig. 24. Each of these later network diagrams represents one particular approach to the construction and is therefore a specific plan rather than an analysis of the problem. This is an example of jumping to a design without carrying out any calculation and should be avoided.

Estimating durations and calculating project times

(Note that this first paragraph is an exact copy of what was said for arrow diagrams, since the determination of activity durations is exactly the same; it is repeated here simply to avoid having to refer back.)

Up to now the network has been drawn without regard to time in order to ensure that the logic is not distorted. It is now necessary to fix durations and as a first approach, an estimate is made for each activity. For convenience the time unit to be used in each project is selected such that it is possible to work in whole numbers of that unit. For example, it is common to use either a week or a month in the case of an outline network for a major project, commonly a day is an appropriate unit for a working network for a building or a computer project, and perhaps an hour for a machine overhaul. It is generally recognised that estimates can only be approximate because of the absence of a direct precedent in the case of project work, and for the immediate discussion a single time estimate is used. There are ways of dealing with variability in activity durations, and this is considered in Chapter 13. The estimate of a duration should be made wherever possible on the basis of prior experience, taking account of any variation in conditions. Activity durations are expressed in terms of elapsed time, as was discussed earlier in this chapter, but some care has to be taken with this. Most activities will only progress during working hours whereas others simply represent the passage of time, e.g. a computer programmer will only work on a project for say 50 hours per week, while the delivery of a new piece of hardware might be expressed in terms of weeks. It is necessary to express both of these activities in the same units, not simply in hours but more specifically working hours. In order to do this a decision must be made on the number of working hours in a week, such as 50 in the case just quoted, and all durations then quoted in that unit; e.g. a delivery of 2 weeks would become 100 hours.

The procedure for calculating project duration in precedence networks is described below and shown in Fig. 25. For each activity it is usual to calculate four separate dates, and insert them on the precedence diagram in the positions indicated in Fig. 26. These 'activity dates' have the same meanings and values as those calculated for the identical project using an arrow diagram. 'Earliest start date' (ESD), which is the earliest date that work can start on that activity, is shown in the top left-hand corner of the box. 'Earliest finish date' (EFD), which is the earliest date by which work on that activity can be complete, is shown in the top right-hand corner of the box. 'Latest start date' (LSD), which is the latest date by which the activity must be started if the project is to be completed on time, is shown in the bottom left-hand corner of

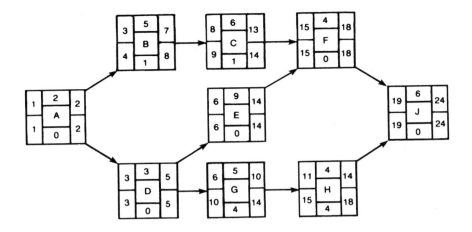

Fig. 25. Precedence diagram — calculation of earliest and latest start and finish dates and float

the box. 'Latest finish date' (LFD), which is the latest date on which work on the activity can be finished if the project is to be completed on time, is shown in the bottom right-hand corner of the box.

Unfortunately there is a slight complication that must be clarified in relation to start and finish dates, and that is the fact that in terms of work progress the start and finish of any one day are one day apart. This means that if an activity has a duration of one day it will start and finish on the same day. Start dates are therefore always shown as being one day number higher than the finish date of the preceding activity. It does make sense that on the start date of an activity it will normally be expected that work will commence in the morning, and on the finish date work will end in the evening. Fig. 27 shows in the form of a bar chart the start and finish dates for an activity, the event times, and also the total float; this should help to make clear how these different terms relate to each other. While most planners will work on this basis it is not universally true, especially where time units are measured in hours or

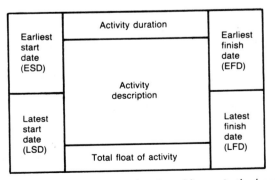

Fig. 26. Precedence diagram activity — explanation of figures in the boxes

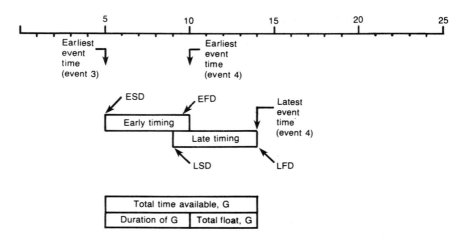

Fig. 27. Bar chart illustration of event times, start dates and float for activity G in the network of Fig. 19 (arrow) and Fig. 25 (precedence)

minutes, because in these cases the activity will begin or end at a single point in time, not on 'hour one' without specifying whether 'hour one' means noon or one o'clock. Readers should not despair at these minor complications; most networks are now calculated by computer, and each package will clearly state which convention is being used.

There are other conventions for the actual layout of the box but the one used here is fairly common. The activity description is usually written into the centre of the box; for convenience and simplicity in Fig. 25 a single code letter is used as a description, but this is not really good practice since it makes reading the network more difficult if a coded list of activities has to be consulted. It is much better if a description, albeit abbreviated, is given. The duration is given in the middle top slot and the total float in the middle bottom slot. The meaning of the phrase 'total float' is discussed later and is the same as in arrow diagrams.

Calculation of start and finish dates Calculation can now proceed. In order to permit a comparison of the two methods of drawing networks the logic used in the arrow diagram of Fig. 19 is now shown in precedence form in Fig. 25. Activity A has to be completed before both B and D, and arrows are shown to indicate this; similarly activity C follows B, and activities E and G follow D, and so on through the network. Activity A starts on day 1 and is finished on day 2, meaning that activities B and D can each start at earliest on day 3. Given this information it is then possible to insert the earliest finish date for activity B as being day 7; this is calculated by taking the earliest start date of job B, plus the duration which is 5 days, and then deducting 1 day. Remember that the day is deducted to allow for the fact that we are using the beginning of start dates but the end of finish dates; the work on activity B will at earliest occupy the 5 days numbered 3, 4, 5, 6 and 7. It is now possible to progress through the network calculating earliest start and finish dates for all activities; in doing this it is necessary to have worked out the earliest start and finish dates for all preceding activities. Care must be taken, however, when faced with an activity which has two immediate predecessors, for example in the case of activity F. We can see from the network that the EFD of activity C is day 13, and the EFD of activity E is day 14. Since it is essential that both C and E are complete before F commences, this cannot be until day 15. A similar situation arises in the case of activity J which must wait until both F and H are completed, i.e. at the end of day 18 not day 14, so that work cannot commence on activity J until the next day, day 19. These calculations lead us to a minimum project duration of 24 days, which is the EFD of activity J. The next step is to reverse the process to work out the latest start and finish dates of every activity in the network.

If we are to complete the project in the minimum possible time it is necessary that the latest finish date of the last activity is equal to its earliest finish date, i.e. day 24 in the network of Fig. 25; this number is then entered in the bottom right-hand corner of the box for activity J. If activity J has a LFD of day 24 it must start soon enough to allow for the duration of 6 days, and hence the latest start date (LSD) is given by LFD of activity J minus the duration of activity J, plus 1 day. (Again it has been necessary to add 1 to get us to the beginning of the next day.) The LSD is then entered in the bottom left-hand box of activity J. We then continue to work back through the network; in the case of activities F and H the LFD will be one day before the LSD of J, and their respective LSDs will be calculated by deducting durations and then adding 1 day. Care has to be taken when calculating the latest dates for an activity which has two or more successors, for example in activity D in Fig. 25. We can see that activity E has a LSD of day 6, and activity G has a LSD of day 10. We must remember to leave enough time for all subsequent activities to be completed and therefore must ensure that D is done in time to allow the longest subsequent path to be completed; in this case that means the path through E, F and J. Consequently the LFD of activity D must be the day before the LSD of E, i.e. day 5. There is only one other activity which has two successors, A, and this is handled in a similar way. It is important to remember that in order to work out the latest dates for any activity it is first necessary to have written in the latest start and finish dates for all following activities. Once the calculations are complete it will be seen that the earliest and latest start dates for the first activity are coincident. If this is not the case then a mistake has been made in the calculations; we hence have an in-built check of the arithmetic.

One of the problems of learning network calculation methods is that the description is much more difficult to understand than the calculation itself. It is of comfort to know that in very nearly all practical cases computers are used to carry out the calculations, but as always with computer outputs it is a good idea to understand what has been done.

The critical path in a precedence diagram   It is worth reflecting that the total project duration is determined by the sequence of activities which requires the longest time, and it is therefore the longest path through the network, not the shortest. Comparison of ESD and LSD at all events will show that there are several at which the two values are the same, and furthermore they form a continuous path through the network. This means that at each of these events there is no time to spare since the earliest possible date coincides with the latest allowable date, and the project is critically dependent upon these activities being started on that date. The path through these critical activities is called the 'critical path', a phrase which applied initially to the whole subject area and became known as the 'critical path method' (usually shortened to CPM). In more recent times it has been realised that while the critical path still features strongly in the management of projects it no longer covers the whole area which is now known as project management. (In recent years the acronym CPM has usually come to mean the arrow diagram method, but the term 'critical path' is still used to describe the sequence of critical jobs in precedence networks.)

The critical path must not be brushed aside since it gives us a very useful means of focusing attention in a project.

- Although it is the longest path it gives the shortest possible duration of the project. In practice it is seldom possible to complete a project in this time because of limitations on the availability of resources, unforeseen delays and so on.
- All critical activities must be complete on time if the project is to be finished on time. If it is required to accelerate project completion then the critical path must be shortened; note however that reducing the length of the critical path is not in itself a guarantee of early project completion (see Chapter 8 on project acceleration.)
- The critical path does serve to concentrate the minds of those who are responsible for activities on that path, ranging from the client assessing needs for a new computer at the start, through the systems analysts and software engineers, and finally to the installation and hand-over.
- All activities on the critical path have zero float (see below for definition of 'float').

Float in precedence diagrams

The preceding section concentrated on critical activities, i.e. those with no time to spare. There are however other activities in a network which are not critical and do have more than enough time for their completion, e.g. activity G in Fig. 25. This spare time is referred to as 'float', which is a useful concept and can be explained as follows. It is the difference between the time available for an activity and the time needed to complete it. There are two main types of float which are commonly used, 'total float' and 'free float' and these are calculated in the following way.

The total float of an activity is given by the difference between the ESD and LSD for that activity. In the case of activity G in Fig. 25 this would be $10 - 6 = 4$. The free float on an activity is defined slightly differently, and actually forms part of the total float. Free float is given by the EFD at the end of the activity being considered deducted from the ESD of the following activity, minus 1 day. In this case we have not taken the maximum available time but the time available without causing any delay to a subsequent activity. Referring again to Fig. 25 this would be the equation for activity H: $19 - 14 - 1 = 4$. As it happens the total float for activity H is also 4 days, but this is not a coincidence; on a non-critical string of activities where it rejoins the critical path the total float on activities in that string will appear as free float on the last activity in the string. Total float is usually written into the lower section of the activity box, as shown in Fig. 26. As an aid to understanding event times, activity dates and float, these are set out on a bar chart for one activity in Fig. 27 (page 78).

All this may appear to many readers as an over-indulgence in the manipulation of numbers, but with a little practice it is not difficult to become familiar with the concepts of total and free float and the difference between them. The good news is that it is rarely necessary to carry out the calculations by hand, since this is easily done by computer; it is important though to know what they are and how they can be used. The management of time in projects really comes down to the manipulation of activity durations and floats and is an essential part of project management. Several examples of this will be discussed in later sections and chapters, but before reading these it is advisable to become familiar with the terms and calculations by spending some time on the exercises given in the appendices.

Other forms of float in both arrow and precedence diagrams

Some readers may have heard of the terms 'negative float' and 'independent float', but these are not in widespread use. Negative float arises only when a project completion date is imposed which is earlier than the earliest date given by the network. The calculation of total float under these circumstances will have a negative value for critical activities. The idea behind this is to concentrate the mind on the critical path and seek to eliminate the negative float. It is probably a better idea to replan the project such that the earliest completion date is acceptable, and then strive to meet that date. Independent float is even less frequently used because it seldom occurs, but it does have value; it forms part of free float. Earlier it was stated that free float was that part of total float which did not affect a subsequent activity; it can however be used up if a preceding activity runs late. The feature of independent float is that it neither affects a subsequent activity nor can it be affected by a preceding activity. It is worked out by assuming that the preceding activity runs as late as possible and the following one starts as early as possible; if there is still more time available than is needed for the duration then the difference is independent float.

The use of float

Float is one of the tools available to project managers. After the foregoing lengthy descriptions of start and finish dates, event times and floats and the somewhat tedious calculations we now come to the important part, which relates to the use of float and perhaps helps to explain why the calculations are necessary. There are many uses, some of them covered here but others in later chapters, perhaps most importantly in relation to the use of resources in Chapter 9. Some of the simpler uses are as follows. First, to extend the duration of activities which have significant float. Care must be taken to ensure that following activities are not unnecessarily delayed, and this is best done by using free float for this purpose wherever possible. Second, to delay the start of activities with long periods of float. For example it may be desirable to delay work which is weather-dependent until spring or summer. In other cases it may be useful to postpone delivery of delicate items of equipment to minimise risk of damage. There may also be benefit in postponing activities with long float if this defers the payment for such work. In some product development projects it may be beneficial to wait for the newest or latest ideas to be incorporated, and there may be many more ways of taking advantage of the spare time provided by float. Resource manipulation remains one of the most important uses of float.

What often happens however is that float seems to disappear because its use is not controlled, components will arrive a day or two late, someone makes a mistake or just does not turn up for work, possibly decisions are delayed or simply lost in the post. The important message here is that all uses of float should be properly controlled and authorised, or what will happen is that the project will appear to be running smoothly and then suddenly goes wrong; what has happened is that too much float has been used too early in the project and there is not enough left to manage the project properly. This raises an interesting question: who is the owner of float? In many cases the project sponsors or owners will claim this since they are paying for the project and therefore feel they have the right to use float as they wish in order to change decisions, delay parts of the work, add to it or modify it in some other way. In contrast with this the suppliers of project goods or contractors will claim that their price is based on the work as originally envisaged and they need to use the float to plan the work to fit in with their other commitments. There is no single correct answer to this conflict; what can be said is that the problem should be foreseen and an agreement reached as to how float should be managed. Perhaps the best answer is that it should be left to the discretion of the project manager, but this in turn means that his or her responsibilities and authority must be clearly set down first.

Linked bar charts

This form of setting out a project plan has evolved not only from the arrow and precedence diagrams but also from the much older concept of a bar chart, which has the main advantage of giving a good visual representation of what is planned. It combines the ideas of drawing a line to represent each activity with a logic linking system, as shown in Fig. 28, which is based on the same activities and durations as the numerical example used in this chapter for arrow and precedence diagrams. It has become popular as a means of presenting a logic diagram on a computer, since it is possible to 'draw' the activities and their links directly on to the screen, giving an instant picture without the need for further processing. Perhaps the main drawback in this system is the need to calculate the duration time for each activity as it is entered — many planners believe that it is preferable to set down the logic on its own at first. It does also somewhat contravene the principle that was stated earlier that the first step should always be to analyse the project, and then prepare a working plan of action as a later step.

The procedure for preparing a linked bar chart as in Fig. 28 is to identify the first activity, A, and draw a bar to represent its duration. The question of which activities

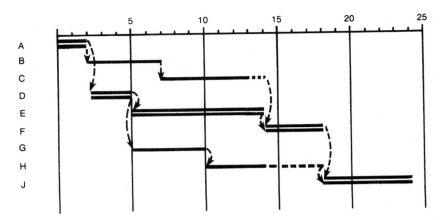

Fig. 28. Linked bar chart for the network in Fig. 19 (arrow) and Fig. 25 (precedence)

can begin once the first is complete will give answers B and D; bars are now drawn to time scale starting at the completion of A, and with arrows to link the dependency of B and D on A. In the case of activity C there is 1 day to spare before F can begin, and this is the amount of free float on that activity, and is indicated by a dotted line. Similarly H can finish 4 days before activity J can start, i.e. 4 days of free float. This particular form of presentation does not directly show the total float on other activities, but it can be deduced by looking at the sequence of activities which ends with a dotted line. Another way to show where total float exists is to indicate the critical activities with a different type of line, such as the double line used here, or in a different colour. (The latter can cause problems if you are using a monochrome computer screen.)

Overlapping activities — arrow diagrams

In some situations, especially where there is some repetition of work it is desirable to permit some overlap of activities, and although the two forms of diagram used so far do not allow this, they can be adapted to do so. Consider the very simple example of building a long garden wall, in which there are really only four activities

(*a*) excavate for foundations (shown in figures as 'Exc')
(*b*) concrete foundations (shown in figures as 'Conc')
(*c*) build brickwork (shown in figures as 'Brick')
(*d*) fix stone coping (shown in figures as 'Coping').

With what we have discussed so far the logic for this would be as shown in Fig. 29 in arrow format and as in Fig. 30 in precedence format. If we now wish to permit some overlap of these activities, which is what might be done in practice, we proceed as follows. In the case of the arrow diagram it is necessary to split each activity into a number of sections, each to represent a defined length of wall, and then show these as separate arrows as in Fig. 31. In order to maintain the correct logic it is necessary to incorporate a number of dummy arrows, and the diagram now shows that the

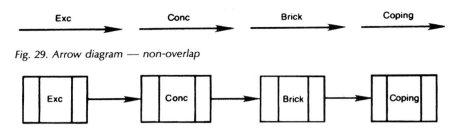

Fig. 29. Arrow diagram — non-overlap

Fig. 30. Precedence diagram — non-overlap

concreting of the foundation over section A can commence as soon as the excavation of that section is complete; it goes on to state that the brickwork on that same section can begin immediately thereafter, and so on. (We shall ignore for the moment that it would normally be the case that a delay of perhaps 3 days would be required between casting concrete and building brickwork on it, but we will return to this shortly.) It is necessary to point out another small difficulty, at the point where the concreting of section A is complete; it would at first seem to be possible to show the relationship as in Fig. 32 by inserting a dummy from the end of 'excavate B' to the start of 'concrete B'. We already have the sequence that 'brickwork A' follows 'concrete A', but as it is now drawn the logic also states that 'brickwork A' cannot commence until after the completion of 'excavate B' (because it follows on from the dummy), but this is not true since there is no physical dependency between the two. Other similar errors of logic arise at events 5, 6 and 7. It is possible to overcome this problem of false logic in the diagram by inserting another dummy between the finish of 'concrete A' and the start of 'concrete B', as in Fig. 31, which also shows similar dummies at a number of other nodes. Reference back to the earlier discussion of dummies may help to clarify this logic.

The resulting arrow diagram for this very simple project is rather cumbersome, and it is in this situation that it is perhaps easier to make use of precedence diagrams.

Overlapping activities — in precedence diagrams

The approach here is quite different from that in the case of arrow diagrams, and is best explained by reference to Fig. 33 where there is no need to split each of the four main activities into three separate sections A, B and C. The procedure in this case is to insert a series of lead and lag arrows connecting the boxes which represent the activities. The first of these, a 'start-to-start lead' runs from the start of the 'excavation' box to the start of the 'concrete' box, and is given a duration equal to the necessary lead time for the work, say 2 days. In a similar way there is a 'finish-to-finish lag' arrow which runs from the finish of 'excavate' to the finish of 'concrete', say 1 day, which specifies that concrete work cannot be completed until 1 day after excavation is finished. Other lead and lag arrows are inserted between the remaining activities, and it is clear that this presents a much more compact network than the arrow diagram in Fig. 31. It also provides the means of overcoming the difficulty which was ignored in the arrow diagram, i.e. the delay needed between concreting

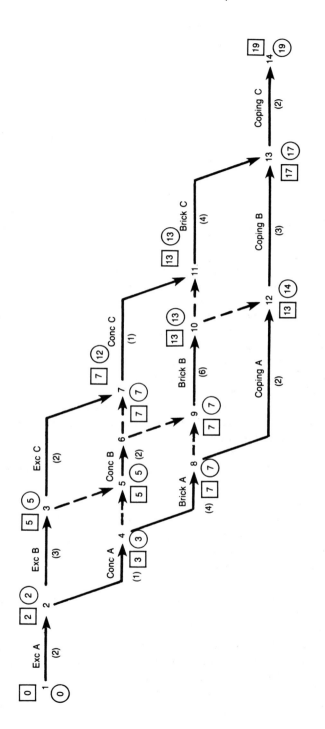

Fig. 31. Overlapping activities with an arrow diagram — use of dummy activities to avoid false dependencies

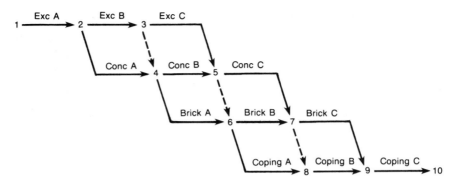

Fig. 32. Erroneous arrow diagram for overlapping activities, implying that brickwork A must follow excavate B

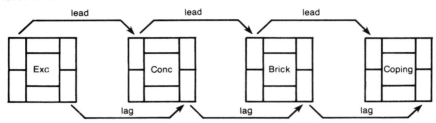

Fig. 33. Precedence diagram for overlapping activities, showing lead and lag arrows

and bricklaying to allow the concrete to set. In the precedence diagram it is simply a matter of ensuring that both the lead and lag times between concrete and brickwork are at least equal to the curing period. The solution in the arrow diagram would be to insert another activity, that of 'concrete curing'.

The calculation of the earliest and latest start and finish dates when lead and lag times are included is slightly more complex, but again is perhaps more difficult to explain than to actually carry out. Rather than expressing it in terms of a series of formulae the best idea is perhaps to work through the same example, now with numbers inserted as in Fig. 34, and using the same acronyms as before; earliest start date (ESD), earliest finish date (EFD), latest start date (LSD), latest finish date (LFD), and abbreviations for the jobs excavate (exc), concrete (con), brickwork (bwk), coping (cop).

First carry out the forward pass (i.e. work from the start of the project through to the finish).

- ESD (exc) is on day 1 of the project.
- ESD (con) is on day 1 plus the lead time from (exc) to (con), i.e. on day 3, with corresponding calculations for ESD (bwk) and ESD (cop).
- EFD (exc) is on day 7 of the project, i.e. its duration after the project start.
- EFD (con) is the greater of two numbers, either the ESD (con), minus 1, plus the duration of (con), or it is the EFD (exc) plus the lag time from (exc) to (con). In actual numbers this is the greater of $3 - 1 + 4 = 6$, or $7 + 1 = 8$, hence we select 8.

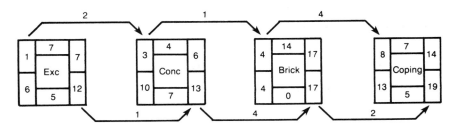

Fig. 34. Precedence diagram for overlapping activities, showing lead and lag times and calculating earliest and latest start and finish dates

- EFD (bwk) is similarly the greater of $4 - 1 + 14 = 17$, or $8 + 4 = 12$, and we select 17.
- EFD (cop) can then be seen to be on day 19, i.e. 2 days after the completion of brickwork.

We now have the minimum duration for the project as 19 days, and can calculate the latest start and finish dates by a similar process, on a backward pass (i.e. commencing the calculation from the finish of the project).

- LFD (cop) must be day 19 if we are to complete in minimum time.
- LFD (bwk) is LFD (cop) minus the lag time from (bwk) to (cop), i.e. $19 - 2 = 17$.
- LSD (cop) is LFD (cop), minus the duration of (cop), plus 1 day, i.e. $19 - 7 + 1 = 13$.
- LSD (bwk) is the smaller of two numbers, either the LSD (cop) minus the lead between (bwk) and (cop), or the LFD (bwk) minus the duration of (bwk) plus 1 day. In terms of actual numbers this would be the smaller of $13 - 4 = 9$, or $17 - 14 + 1 = 4$, and we select 4. A similar process gives the latest start and finish dates for (con) and (exc).

It can be seen that the LSD (exc) is on day 1 which is the same as the ESD (exc), confirming the arithmetic. Note also the use of the beginning of the day for the start dates and the end of the day for finish dates due to actually starting work at the beginning of a day and finishing work at the end of a day (see earlier discussion on start and finish dates on page 72).

It is perhaps in connection with the use of lead and lag times that precedence diagrams have their biggest advantage over arrow diagrams, but they do carry one risk. This is the fact that simply because a lead time has elapsed it cannot be guaranteed that enough of the first activity has been completed to allow the second to start.

LAST WORD ON TIME

Much detailed attention has been given here to the calculations of time, and this is because of the new concepts introduced by network planning and control methods. It does not imply that time is more important than cost or quality in project management. Time is however a rather slippery commodity, it passes inexorably onward and cannot be made to go in reverse, or even to stand still. Any time lost is gone forever and cannot be recovered; it is therefore well justified to give it special detailed consideration.

Many authors have devoted time to descriptions of network methods, and readers who require more practice of the calculations would be well advised to consult some of their books (Ahuja et al., 1994; Love, 1989; Spinner, 1992; Pilcher, 1976). There are many others.

6 The management of cost

Throughout industry there has been a very strong tendency to measure everything in terms of cost, largely because there is inevitably a great deal of information which is measured in these terms; wages, purchases, rent, taxes, sales, profit and so on. Added to this is the fact that in the UK the accountant has become in many cases the leader of control activity and it is mainly in terms of money that accountants think. In recent years it has become fashionable to talk in terms of 'the bottom line' as being the dominant factor in any undertaking, taking precedence over such aspects as value, time, environment and quality. Project management has done much to lead the way in taking a wider view of what we do, and looks carefully at time, cost, quality, and the way they interact, as well as the costs and benefits of projects to the community. Having said all this it is essential to look at cost and cost management, and these are the subject of this present chapter, in which an examination is made of the planning and control of costs *per se*, as well as the use of costs as a means of measuring progress on a project. There are at least three distinct aspects of cost in relation to the management of projects, and it is important to distinguish between them.

PROJECT TOTAL COST

Clearly the economic viability of a project will depend upon its total cost, and at the early stage of any project an estimate of this has to be made. This can usually only be approximate until the project has been designed, but a reasonable estimate is needed at the time that the project appraisal is carried out, since it will be a major factor in the 'go/no go' decision at that time. The total project cost will be the sum of its parts and these should be spelt out in reasonable detail. One of the discussions which has to take place at the early stage of any project is the final determination of the scope of work to be included. If the project appears to exceed the funds that can be made available for it then the scope of work will have to be reduced or modified in some way. Such modification may be in terms of a reduction in the standard of work being specified, as is described more fully in Chapter 8. However, as is emphasised in the discussions of quality in this book the quality standard once specified should not be negotiable, and the temptation to reduce cost by a relaxation of quality achieved must be resisted.

The total project cost figure will go through a number of changes as the project proceeds. The original estimate will be based on a simple outline design, which will include many unknowns, and it is therefore usual to add on a 'contingency' allowance, which is really an ignorance allowance. The original estimate only may be used as the basis of a decision to proceed to a more detailed design, which will then permit a more accurate estimate to be made, in which the contingency allowance is reduced. A further review may then be made, allowing a decision to proceed, followed by the seeking of formal prices for the execution of the project. This is actually a rather complex process which can be carried out in a number of ways depending upon the nature of the project and the industry which is carrying out the work, as is discussed in Chapter 11.

Once a tender for the work has been accepted and a contract placed then the figure that has been referred to as a cost really becomes a price; it is the price that will be paid by the client to the supplier, adjusted for any changes that may be made to the work scope or detail during the project. From the client's point of view it still should be thought of as a cost, as it is a liability which is being incurred, and indeed it should still include a contingency allowance. The size of this allowance will depend upon the nature of the work. If the project comprises the supply and installation of standard equipment, then there should be only a very low figure, perhaps of less than 1%. On the other hand work which is below ground level may be greatly influenced by conditions which cannot be foreseen until the ground is excavated as part of the contract; construction projects therefore may have contingency allowances of up to 10%. The size of the allowance may be reduced as the project proceeds and the areas of uncertainty are more limited. The methods of working out total project costs are discussed more fully later.

Life-cycle cost

The appraisal of projects is dealt with more fully in Chapter 12, but it is opportune to mention the concept of life-cycle cost at this stage, in the context of total project cost. It can be argued that the comparison of alternative designs for a project should not be based simply on the initial capital cost, but should take account of operating, maintenance and decommissioning costs over its whole working life, as well as the revenues generated. This can then take into account the total net cash flows of the project, and also the time value of money and the impact of inflation. This does indeed complicate the use of money as a measure of project value, which is why it is mentioned at this stage of the book; it gives another reason why money may not be the best yardstick against which projects are measured.

COST AS A MEASURE OF PROGRESS

The ready availability of cost information means that it is often used as the basis of measuring progress on a project, and this is looked at more fully in Chapter 3 which shows that in many ways this can lead to problems. The main difficulty is that the fact that money has been expended does not guarantee that value has been created.

In most projects it is the creation of value that is being sought, and it should therefore be the assessment of value created which is used as the basis of measurement of progress. Having said this it is still the case that many project control systems will continue to use cost information for control purposes, and it is almost inevitable that the same data will be used to measure actual progress. This then brings discussion to the third aspect of cost management.

COST CONTROL

The interpretation given to this phrase here is the measurement and control of costs which are incurred in the execution of each item of work as it is carried out. This is best illustrated by some simple examples. An architect undertakes a commission to provide detailed construction drawings for a new building, and the fee to be paid is defined in relation to the contract value of the completed building. The architect's practice seeks to contain its own costs within that fee, and must have its own targets for the cost of preparing the drawings. A regular comparison of cost and value will indicate to the firm how its own economic performance is working out, and will permit changes to be made if these are required. In a very similar way a building contractor will wish to know whether the excavation, concreting, brickwork and so on are all being efficiently carried out, so that if there is any inefficient work apparent it can be investigated. Similar controls will apply in computer software companies who have contracted to complete the design and installation of an information system, and also for the company which has to supply and install the hardware; computer companies will have to plan expenditure on a project and compare their actual expenditure with this plan at various stages, in order to operate successfully.

COST DATA

Before looking in more detail at the means of estimating project costs either in terms of the client's interest in total price for the project work, or the supplier's itemised costs, it is useful to examine some of the ways in which cost data will be handled. One of the most useful forms of presenting data is in the so-called S-curve, which plots the cumulative cost of a project against a base of time. This is illustrated in the following simplified example, with data as in Table 4 and the graphical plot in Fig. 35, for a one-year project. The concept of S-curves was introduced in Chapter 3 and it may be useful to read that section again before proceeding (*see* page 36).

Note that the rate of cost build-up is low at the start of the year, accelerates to a maximum rate in June and July, and then slows down again to the end of the project. This is typical of most projects, although each will be different in actual shape, and may include some large jumps as in the case of acquisition of major plant items. It is almost certain however that the early and late months will show a reduced rate of cost accumulation.

Table 4. Budget cost data for typical one-year project

Month	Cost in month: £	Cumulative cost at end of month: £
January	500	500
February	1000	1500
March	2000	3500
April	2500	6000
May	3500	9500
June	5000	14 500
July	5000	19 500
August	4000	23 500
September	3000	26 500
October	2000	28 500
November	1000	29 500
December	500	30 000
Year total	**30 000**	**30 000**

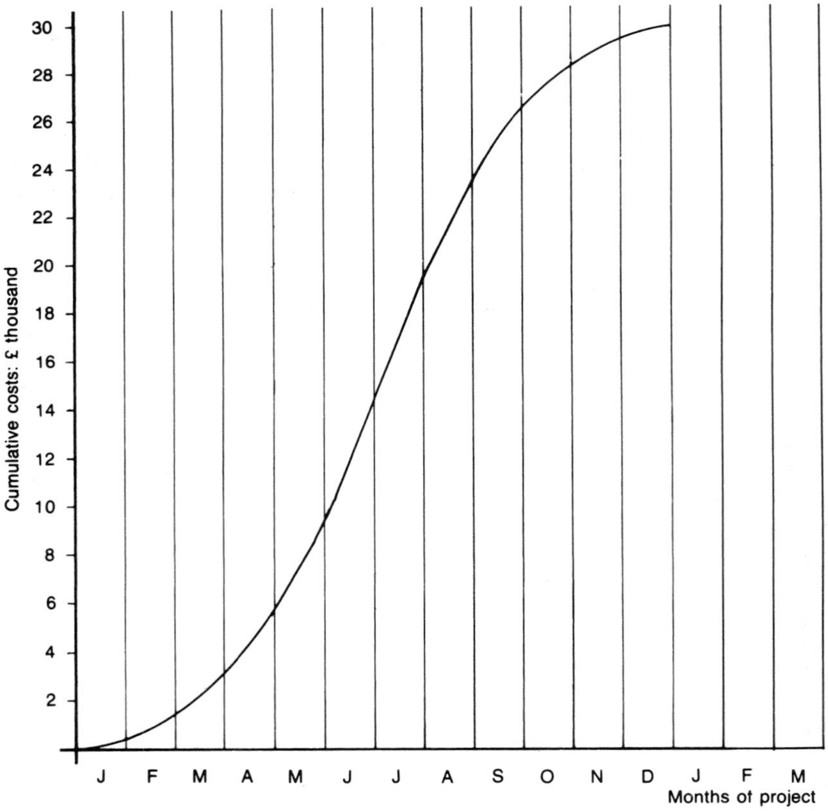

Fig. 35. Cumulative cost curve based on budget costs at planned time

Cost figures used in preparing an S-curve

The next thing to look at is the nature of the cost figures that are used to prepare an S-curve, and the way that they may subsequently be used. Consider the simple case where an in-house project is being undertaken to set up a new production facility, involving some site preparation, delivery and installation of equipment in a number of stages, then its testing and commissioning. There is no complication here of the timing of payments to contractors, and it is assumed that everything has to be paid for as items arrive and work is done. The S-curve of Fig. 35 then could be used at the outset to indicate the cash flow for the project, based on an agreed time-plan for the work. If the project now proceeds it is likely that some variation from the plan will take place, either in terms of the timing of the work or in terms of the cost of each stage, or both. For project control purposes it is not sufficient to plot the actual cash flow on a month-by-month basis, since it is not possible to differentiate between variations which are due to timing changes and those which are due to changes in the cost of the work done. In order to illustrate this point consider Fig. 36 in which it is now assumed that the data used in Table 4 represent the budget cost for the work of the project at its planned timing. A second S-curve is drawn as work progresses, plotting the cumulative total of the budget cost (the planned cost rather than the actual cost) of items completed. This curve shows the budget costs of the work as it is actually completed, and in this instance indicates the same total cost but spread out over a longer period (Table 5), and the horizontal difference between the two lines is a measure of the extent to which the project is behind its time schedule. It can be seen from Fig 36 that there was no time slippage for the first three months of the project, but after that time there was a steady increase in the project delay. Since the cost

Table 5. Budget costs of work completed

Month	Budget cost in month: £	Cumulative cost at end of month: £
January	500	500
February	1000	1500
March	2000	3500
April	2000	5500
May	2500	8000
June	3000	11 000
July	4000	15 000
August	4000	19 000
September	3000	22 000
October	2500	24 500
November	2000	26 500
December	1500	28 000
January	1000	29 000
February	500	29 500
March	500	30 000

figures used are the original budget figures, this comparison of timing is not distorted by any variations which may have occurred in the costs of individual items. This diagram also gives a measure of the expenditure that would be necessary to bring the project back onto its target completion date. For example, at the end of September it would be necessary to complete nearly £8000 worth of work in 2 months at a time when the project is about 70% complete and so has reduced scope for acceleration. It is nearly always necessary to leave about a month at the end of such a project for finishing off the minor details, and it is a mistake to expect to be able to use this last period to catch up on lost time.

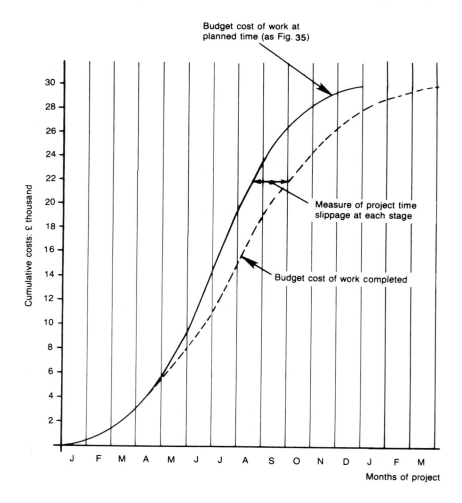

Fig. 36. Comparison of planned and actual time based on budget costs

TOTAL PROJECT COST CONTROL

It is of course important to know whether the costs are being contained within the budget, and this can be checked simply by comparing actual and budget cost for each item as it is completed. There is however the possibility that there is a general cost trend which should be recognised as soon as possible, and this must be based on the work completed to date. It can be illustrated by reference to Fig. 37 which shows a comparison of the same curve as shown in Fig 36 representing the budget cost of work completed (Table 5), and a new one which plots the actual cost of work completed (Table 6). For a true cost comparison to be made it is necessary for these two cumulative cost curves to be drawn to the same time-scale so that they both therefore relate to the actual time at which the work was done.

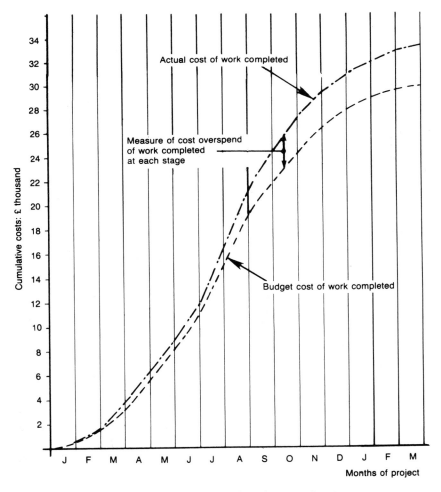

Fig. 37. Comparison of budget and actual costs based on actual timing

Table 6. Actual costs of work completed

Month	Actual cost in month: £	Cumulative actual cost at end of month: £
January	500	500
February	1000	1500
March	2500	4000
April	2500	6500
May	2500	9000
June	3000	12 000
July	4500	16 500
August	5000	21 500
September	3000	24 500
October	3000	27 500
November	2000	29 500
December	1500	31 000
January	1000	32 000
February	1000	33 000
March	500	33 500

Note that the extent to which the costs have over-run above budget is measured by the vertical distance between the two lines of Fig 37; it shows that the variance was fairly small until July, but thereafter a steady cost rise is shown. If this diagram was being used in a project to monitor the cost of work it would be effectively the half-way stage before the variances showed up, leaving half of the project in which to seek to prevent further losses and try to recover from the earlier ones.

Caution in the use of S-curves in project control

There are several reasons why caution must be exercised in the interpretation of cumulative cost curves as a means of exerting control over a project. The first of these is that a single diagram such as Fig. 37 will at any stage of the project, say at the end of November, give an indication of the difference between actual cost and budget (i.e. the cost variance) of the total work completed. It will not indicate where the variance has occurred on the project and may indeed hide variances which offset each other on a 'swings and roundabouts' basis. In order to maintain real control over the work of a project it is necessary to look at costs in much more detail, as is done later in this chapter. Another reason for caution is that the cost curves shown as the actual cost of work complete are likely to be based on data from wages sheets and purchase invoices; the budget cost of work completed is however probably based on notional figures allocated to each work item or package which has been totally completed, and this may ignore the calculation for work in progress. Hence the two lines in Fig 37 may not be comparing like with like, unless adjustments are made to allow for such differences. A third potential problem arises where a priced bid for a contract is used as the value to be assigned to each item of work; in some cases a supplier may seek to spread overhead and profit in a non-uniform way, and this could lead to a positive or negative difference between the price to be paid for work and its real value. There is a fourth factor to be wary of and that is the fact that a satisfactory comparison of both cost and timing of work may hide the fact that work on the critical path is not progressing fast enough (see Chapter 3).

Notwithstanding these criticisms of the use of S-curves of total project costs at each stage, they do give a general picture of progress which can be useful for top-level reporting. When they show major discrepancies it is certain that serious problems have arisen, and further investigation is required. The problem is that they may at times disguise lesser difficulties, and the safest procedure may be to take note of them when they show the possibility of problems, but also to look more carefully when no trouble is evident. Several other similar forms of diagram have been proposed and used (Turner, 1993) devoting considerable time to this important subject, including methods of using trends throughout a project to predict what the ultimate project outcome will be. These can be very valuable in the case of long-term projects where there is still time to make changes to methods of working which have been found to be uneconomic. There is a risk of confusing the different terms used in various books, but careful reading should make the meaning clear, and it is preferable to use the full title of each cost curve rather than an abbreviation. For the benefit of readers who may have come across similar terms the following definitions may be useful. The figures are normally presented in the form of a cumulative S-curve, as described earlier.

'Budget cost of work planned' is the cost at each stage of the project if all goes well and the work is completed on budget and on time. Other books call this either 'baseline cost of work scheduled', 'budgeted cost of work scheduled' (both abbreviated to BCWS), or 'planned cost of work scheduled' (PCWS).

'Budget cost of work completed' is the cost at each stage of the project if work is costed at the original budget rates but at the time it was actually completed, rather than at the planned time. Other books have referred to this measure as 'baseline cost of work completed' (BCWC), or 'budget cost of work performed' (BCWP).

'Actual cost of work completed' is the cost pattern obtained by taking the actual cost of work at the time it was actually completed. This phrase is used by many project managers and abbreviated to (ACWC), but others refer to 'actual cost of work performed' (ACWP). It is easy to see why the use of initial letters alone can lead to confusion.

CALCULATION OF INITIAL TOTAL PROJECT COST

At the stage of the initial concept of a project it will be necessary to base a total project estimate on the general experience of the particular industry, and it will be found that most industries have their own special practices. For example in the field of power generation it will be known by design engineers that the capital cost of a new power station of a particular type can be estimated in terms of £ per megawatt of installed generating capacity. Similarly the cost of a new building can be estimated by quantity surveyors in terms of £ per square metre, within a range to cover the standard of specification of a building. Each industry has its own rule of thumb for making rough estimates which can be used at the initial concept stage, but in order to make a proper project appraisal it is necessary to refine this estimate by carrying out enough design work to allow cost engineers and surveyors to produce a more accurate figure. In most industries it will be normal to make a decision on whether or not to proceed, depending on the investment required. In some cases the decision may be determined by other considerations, perhaps safety or the requirement to meet standards or regulations; it will still be normal practice to make a project cost estimate even where this is not a determining factor. It would certainly not be good project management practice to embark on a project with an open-ended financial liability, unless it was an extreme emergency.

Once a project is past the appraisal and approval stage and has become a 'live' project there will be a need for a fully detailed cost estimate and cost plan. If the work is to be executed by a contractor this will be in the form of a bid, which may or may not be in competition. The bid if accepted will become the price for the work and will be incorporated within the contract for the work. Depending on the nature and form of the contract, payment is made in accordance with this to the supplier/contractor (see Chapter 11). Even where there is no competition it would be usual for an estimate to be prepared and used as the basis for a contract. In the case of a construction project the normal practice is to prepare a bill of quantities which sets out the prices of all the individual project items, of which there may be thousands. The combination of a priced bill and a detailed programme will provide the necessary information to work out a cash flow forecast which will enable the client company to plan the funding of the work.

There may be a temptation for the supplier or contractor to use the same information base for the control of the building process, but this can be seen to have dangers. It is probably preferable for the contractor to base the cost targets on an operational basis rather than the traditional trades basis, and link it to the sources of cost. This is a somewhat specialist subject which has been discussed in great detail by such authors as Pilcher.

Work packages

In the case of non-building work there tends to be a different approach to the calculation of the budget cost for the project, and this is by using work packages. This concept is included as part of the discussion of work breakdown structure (WBS) in Chapter 9. Basically the idea is that it is impossible to derive an accurate budget cost for a project as a whole from rough rule-of-thumb calculations such as £ per megawatt of installed power generating capacity; the work is broken down into a number of work packages each of which is usually confined to an item of plant and/or to an area of specialist work. For example one work package in a power station might comprise the pumping system for cooling water, another the control system, a third the switchgear and so on. The level to which work is broken down will depend in part on the size and complexity of each area of work and the extent to which it is complex and unique, as is described in Chapter 9. Heavy mechanical, electrical, chemical and petroleum construction work has developed the use of work packages, and does not normally follow the use of bills of quantities in the way that would be used in the construction of an office block or hospital project. This is largely because the project will probably comprise a relatively small number of large work packages, which are supplied under a design, fabricate and install type of contract, where the specification is based on performance rather than a detailed description of work as in the case of a building. In such heavy plant projects the civil engineering and building work will be relatively small in value and may comprise one or more work packages.

Detailed control of project production costs

Reference was made earlier in this chapter to the control of costs as work progresses, and the way in which actual costs may be looked at to help control the project. It was pointed out that examination of total cost to date may disguise some upward and downward variations in individual items. If real control is to be exerted over all of the processes involved in the project then it is necessary to look at each of them in isolation. This approach may be normal practice in steady-state factory production where inefficiencies in some processes may be identified and corrected in the subsequent operation of those processes. One of the characteristics of projects however is that they are one-off, and there is often little opportunity in any one project to put into practice improvements which have been made earlier in that project. Another feature of projects is that they usually consist of a wide variety of different operations, and to make a full cost analysis of all of them would be very time-consuming and expensive. The best approach therefore has a number of different strategies. First it is not worthwhile trying to fully control in detail all of the operations in a project. The so-called Pareto principle states that in almost any activity approximately 80% of the total cost of an undertaking will be covered by only 20% of the cost headings. In construction work this concentration may be even more pronounced with 90% of the project value being accounted for by 10% of the bill items. This means that a good overall control can be exerted by watching very carefully the 10% of items which represent the most cost. We shall return to these selected items shortly, but first some comments on general control.

General control of detailed production costs

The approach here is to set up systems which encourage economy and discourage waste without the necessity to make detailed records of every item of work on a project. This will then influence not only the 10% work which is high cost, but will also cover the 90% which are small items. It was stated earlier in Chapter 3 that it is important in project-based work to have a high level of skilled supervision, because of the varying nature of production work. This supervision should be able to prevent waste of time and materials on abortive or unnecessary work, or work which is below standard. Another approach which is widely used in building work is the incentive bonus which can cut down wasted time, but may lead to poor quality and waste of materials unless measures to prevent these are built into the incentive scheme. Some projects use the concept of a general project bonus related to performance of the project as a whole in terms of time, cost and performance; this is paid to everyone, but some observers feel that this is too remote from the individual operative to encourage extra effort. One important cause of cost overrun is the loss of materials which may be due to careless waste and this should be controlled by good supervision. On a building site such waste is usually obvious because it is lying around in the site mud, but on other projects may not be so easily seen. Rather more difficult is the material or components which either go missing or never existed except as a faulty entry on an invoice. The way to control such waste is to have rigorous materials checking and accounting systems, and on big projects these are found to be very worthwhile.

Cost control of the important 10% of items

In most project work there will be three main headings of direct cost, i.e. those costs which relate directly to each item of work, usually in proportion to the amount of work done. Let us leave aside overheads for the moment. The three main generators of cost are labour, plant and materials and these have to be taken separately. Most projects will involve all three of these, for example a computer installation will need many days of skilled engineering and technician time, it will require a range of manufacturing equipment and test machines and also will consume component parts and other materials. The construction of a concrete bridge will involve among other things excavation work, piling, shuttering, reinforcement and concrete, which will require various types of skilled labour, the use of plant such as excavators or cranes, and will consume materials including steel, concrete and timber. It is likely that many of the individual tasks included in the largest 10% of items will be repeated over a number of weeks, and it is possible to see where improvements in performance can be made, for example by studying the unit costs of these items week by week. Taking the example of brickwork it is fairly easy to measure the work done each week, calculate labour costs from wages sheets and materials costs from invoices, although the latter may be complicated by variations in stock levels held. This will make possible a weekly calculation of the unit cost of each type of brick wall in terms of £ per square metre, and by comparison with previous weeks any trends can be observed and acted upon where appropriate. There are several books which refer to construction cost planning and control (Cooke and Jepson, 1979; Harris and McCaffer, 1977; Pilcher, 1985).

Standard costs

Comparison with a previous week is not really a good basis for control, since it is possible that the performance in all earlier weeks has been poor, and even if there is improvement it is still not satisfactory. Comparison with bill prices is also not reliable, since some prices may have been modified to take account of overheads. Furthermore the use of bill prices on site has other risks, one of these being that if the actual cost comes out at the bill price there is a view on site that this gives a satisfactory performance, whereas if the actual cost is above bill price the attitude might be 'you can't win them all'; the outcome of these two views is that the project will overall lose money. In a situation where there is a healthy margin between the actual cost and price it removes the incentive to keep performance at a high level and may even put pressure on the bonus scheme. As stated earlier there may be good reasons for spreading overhead charges in a non-uniform way, and larger margins are then necessary on some items. In order to overcome these problems a common practice adopted from manufacturing is to calculate a so-called 'standard cost' which represents what should be achieved by well-motivated competent operatives working steadily under good conditions. Each week's performance is then compared with the standard cost for each item of work and the difference between actual and standard is worked out; it is known as the 'variance' and may be either positive or negative, leading to appropriate action in each case. The procedures of on-site cost control can become fairly complex and burdensome, and are not further discussed in this book; there are several useful references on this subject, including those by Pilcher (1985), Cooke and Jepson (1979), Seeley (1976) and Harris and McCaffer (1977).

Cost planning of a single item of building work

The task of working out the detailed cost of work in the construction industry is usually carried out by a person called an estimator, who may be an engineer, a builder or a quantity surveyor. It is usual to 'build up' a rate for each item of work in a bill of quantities, on the basis of the three main components of labour, plant and materials. The material component is usually fairly simple, as for example in the case of brickwork where there can only be a fixed number of bricks per square metre of wall with a small addition for waste due to cutting, and for the mortar. Labour and plant costs are more difficult to predict because of uncertainties relating to speed of working, whether access is at ground level or from scaffolding and so on. Most contractors will have their own records of performance rates and these can be modified by the judgement of the estimator. A particular problem relates to the use of large plant items such as cranes which may be used for short periods of time over a wide range of tasks. One way to incorporate the cost of cranes is to treat them as an overhead, the cost of which is spread over all the work on the site on some empirical basis. It is possible to compare prices of various construction activities with published data on output rates and prices, e.g. Spon's Price Books. (Davis et al., 1984).

The preparation of contract bids

It is not appropriate to enter into detailed description in this book of the preparation of bids by contractors, but it is of relevance to consider the general form of them. The procedures are very well described in Spon's *Architects' and Builders' Price Book* and their *Civil Engineering Price Book*. These books give not only a description of the estimating work to be done but also a great deal of detailed data on the actual costs that would be appropriate. Hourly costs for labour and plant and unit costs for a wide range of work items are quoted and updated annually. These are based on average working conditions with appropriate notes about possible variation, and it is made clear that they are the direct costs of executing the work, do not include overheads and profit and therefore cannot be regarded as prices which might be expected in a submitted tender. There are also regular reports published in technical journals which give prices based on recent contracts. All of these sources cannot be relied upon alone for the calculation of prices, but they provide a good back-up to an estimator's work, which must take account of the specific conditions of the site in question and the terms of the contract. One important general point that should be made is that the estimator's job is to prepare the closest possible estimate of what the work will actually cost to complete. A second and quite separate step has to be taken in converting this to a bid for a tender submission; an overall management decision has to be taken about the margin to be added for overheads and profit, to cover unforeseen difficulties which cannot be charged, and any addition or deduction which relates to the bidders' keenness to win the contract. A phrase sometimes used in the contracting business is to ask the estimator to 'use a sharp pencil', i.e. to use cost estimates that are on the low side. This is a mistake, as the total cost estimate will almost certainly be low, but the firm's management will not know by how much, and they may make further cuts in order to submit a competitive bid, without realising that they are perhaps then bidding below cost. If such a bid is successful it is clearly bad for the contractor, but will also be bad for the client as the contractor will be in difficulty and will seek extra payments based on claims, or even worse may go into liquidation with the consequent disruption to the project. The most successful projects are those which pay a fair price for good work.

CONTROL OF PROJECT COSTS

An earlier paragraph discussed the overall control of project costs, but there is another more detailed area that has to be considered, namely the control of individual item costs. It may not be obvious that this is a matter of great concern to the project manager, and in fact many project managers are only involved with the client's direct interest, namely what the cost to the client will be. This may be true of projects where there is a client/contractor relationship which isolates the client from individual item cost variations. There may be many projects where no such relationship exists, for example in the case of in-house projects where all costs incurred are met directly by the client, or in the case of 'cost-plus' contracts in which all project costs have to be reimbursed by the client. Another situation of course is that of the contractor or equipment supplier who has to manage the project and to complete it not only on time and up to specification, but also at the minimum cost consistent with these other factors, so as to maximise the profit.

There are many well-tried cost control procedures used in manufacturing industries and some attempts have been made to adopt these in project-based work. There is a major problem however due to the one-off nature of work in projects. Even if good data on actual costs is recorded and points to the need for improvement in performance, it is likely that on any one project there will be no similar work to follow to which improvements can be applied. The best that can be done is to break work down into elements which have more general application, record data and then calculate unit costs of these elements and compare them with similar tasks on other recent projects. It is only realistic to attempt this task for those activities which form part of the 90% of the value of the project, but as stated earlier this does reduce the number of items very significantly. In the case of building projects, cost data can be usefully carried out on common tasks such as bricklaying, plastering, and drainlaying, but the variations that occur locally on ground conditions may mean that it is difficult to make much use of comparisons of costs of excavation work, except in very general terms.

7 Quality management and project performance

This chapter aims to introduce the broad concept of quality in an industrial context and to show how this differs from the popular concept of quality. Quality in project-based industry will also be examined and the subject of how it can be properly managed will be discussed. The historical development of the subject will also be outlined. In order to give it some reality the description is mostly in the context of the construction industry, but this should present no problems for readers more concerned with other industries as there is no use of highly technical terms and the generality of the argument is applicable to virtually all project work. Some non-construction examples are quoted to give relevance to particular points.

WHAT DO WE MEAN BY QUALITY IN PROJECTS?

It is important at the outset to distinguish between the everyday use of the word 'quality' and the more specific meaning we use when talking about the management of quality in projects. The man in the street may often make reference to high-quality products as distinct from medium-quality products (but will seldom refer to low-quality goods). For example, a Mercedes car is often referred to as a 'quality' vehicle compared with simpler cars. What is really meant is that the Mercedes has a higher specification. However both cars should be reliable and should perform up to their own specification, and both should therefore have good quality. In construction projects a 'high-quality' building facing may be polished marble, whereas brick or artificial stone will be seen to be of 'lower quality'. Again the marble should be thought of as having a higher specification, and both materials should be correctly laid or fixed and should perform up to the relevant specification, i.e. each should be of good quality in the sense that we are now using the word. Generally, higher specification will involve higher cost. The standard of product or material that is specified is set by a conscious decision during the design stage, at which the appropriate required standard is set.

Quality in the sense that we will be using the word is not a measure of the standard set at the design stage, but is a comparison between the standard set for a particular item and the performance that is actually achieved. It is therefore a measure of the conformance of the product to the standard set. For example, it is possible to have a building faced with expensive marble in which the workmanship is poor, with the consequence that joints are irregular or out of line, or where there are chipped corners on the blocks — this would be poor quality. By comparison, a wall could be built of brick to correct line and level, with all joints regular and clean — this would represent good quality of construction.

Quality therefore is a comparison between the standard achieved and the standard required and specified. It is sometimes referred to as conformance or compliance with specification. Setting the standard in the first place has to be considered very carefully, as there is little point in setting a standard which is unrealistic, unattainable or unnecessary. What is relevant is the fitness for purpose. For example, to specify that all bricks should be laid within plus or minus 1 mm of correct position would require very accurate and expensive bricks, would be virtually impossible to measure, and would be quite unnecessary as well as being very costly. A figure of 2–5 mm would be more realistic.

The standard specified should always be the minimum acceptable; this does not imply an attempt to 'get away' with inferior work, but means that there is no point in requiring a standard which is in excess of that which is relevant to the case, especially where this would entail additional cost. It is very important to note that it is only in relation to the initial setting of standard that the concept of 'the minimum acceptable' applies; when it comes to checking the quality of work actually done there should be no possibility of accepting work which does not meet that standard. Once a standard has been set and agreed, that standard should not be negotiable; one of the main causes of argument in construction centres around the question of acceptance of inferior work for reasons of expedience. There is no point asking for (and therefore paying for) a standard higher than that required, but at the same time it should always be made clear that a standard lower than the one specified will not be acceptable. In Chapter 8 there is a discussion of the interaction of time, cost and quality, but it should be emphasised again at this point that quality standards which are initially agreed are not thereafter negotiable.

Much of the work done in the management of quality has been applied to manufacturing industry (Scherkenbach, 1993), and this is discussed later. In order to understand the extent to which construction is similar to and different from manufacturing it is necessary to look at both types of industry. This was done in Chapter 2 but the essentials have been repeated below (Baden Hellard, 1993; European Foundation for the Improvement of Living and Working Conditions, 1991; Atkinson, 1995).

Comparison between construction and manufacturing

As stated earlier in Chapter 2 there is a widespread myth that 'if you can manage one type of industry, then you can manage any other', which implies that the management problems of all industries are the same or at least very similar; this is not true, and it is useful to make some comparisons. For the purpose of this chapter, industry is thought of in two broad groups; project-based and manufacturing, although each clearly has further sub-divisions. (For our present purposes we consider process industry along with manufacturing.)

Considering construction as a good example of project work, we saw in Chapter 2 that it is of a form sometimes referred to as 'jobbing', namely one-off products usually made specifically to a customer's order. This is a very important aspect of construction which influences many of the ways in which the industry is managed; it is therefore essential to remember and fully understand this characteristic. By comparison manufacturing or process industries operate on a batch or mass production basis, with the intention and hope of selling the product subsequently to a customer, who may at the time of production be unidentified — it is usually a matter of making for stock. The management of batch or mass production has many differences from construction, and it is useful to examine some of these differences.

Construction

Listed below are the main characteristics of work within the construction industry.

- Construction is usually one-off, and therefore there is no opportunity to progressively learn from one unit of production to the next in order to improve quality; it is absolutely essential to 'get it right first time', as there will rarely be a second time. This phrase is widely used in many industries, but is particularly true in one-off project work. It is sometimes said that project management is largely about doing the right things at the right time in the right way.
- The customer (usually called the 'client' in construction) is known, and is very likely to have an input and an impact during the production process. This can be useful but can also be a nuisance, especially if changes are ordered to the quality specified and which may then cause disruption to smooth work on site.
- There is a very low level of automation, but a high level of personal operative input, i.e. it is a labour-intensive industry. Most tools and equipment are relatively simple and multi-purpose. (It may be thought that a crane is a sophisticated machine, but it is really only a large and strong hand to lift, hold and place items.) It could be said that conformance with standards of workmanship is literally in the hands of the operatives. This implies the need for good training and a high level of supervision, both of which are often inadequate.
- There is a mobile labour force which moves from project to project, often with more loyalty to a type of work or to a district than to a particular employer. This changing labour force can lead to problems related to continuity and hence to quality. There is an opportunity to build a new team for each project, a feature which can have both advantages and disadvantages. Highly skilled workers are frequently lost, but at the same time there is no obligation to persevere with those whose standards are in doubt.
- Real problems can arise with regard to defective work, e.g. it is often difficult or impossible to throw defects away; it may be necessary to spend much time and money in removing defective work, with consequent delays to the project. (It takes a great deal longer to remove concrete than it does to pour it in the first place.)

- The normal practice of separating the responsibility for design and site construction can inhibit the implementation of good quality; the two teams frequently do not know each other, and communication of detail is more difficult, with no shared culture of quality. There will often be many independent sub-contractors who do not have experience of working together.
- The life-span of a construction project is long, and it is likely that it will undergo modification at various times. It is difficult to foresee what quality demands may arise during this long period, and initial faults may only come to light after many years. In addition, major works in the past were generally over-designed in order to provide a high factor of safety, but modern analytical design methods have permitted 'margins of ignorance' to be reduced. A good illustration of this is that many bridges built more than a century ago are still in use carrying loads far in excess of those that were envisaged at the time of design.
- Any building will incorporate materials and plant items which have been produced under well-controlled factory conditions, but are then subjected to site handling, storage and installation, much of which is not formally controlled, with the consequent risk of damage or other fault.
- Formal testing of on-site work is not universal, and often only takes place when a fault is suspected.
- The recording of faults is not rigorous, and much of the consideration of errors is associated with the apportionment of blame, rather than seeking to prevent a similar fault on subsequent projects.

Manufacturing/process industry

Below are listed the main characteristics of the manufacturing/process industry.

- There is an opportunity to gradually evolve a product or material, learning by immediate prior experience.
- The customer is usually unknown, and cannot individually be asked what is wanted; it is hence necessary to anticipate needs and demands, perhaps by market survey.
- Often there is a high level of automation, i.e. a capital-intensive process, using very specialised equipment. It is important for such equipment to be kept in use, but it is possible to build into the machinery the appropriate quality controls.
- The labour force is relatively static, and can undertake extensive training programmes, e.g. in quality procedures. There is one disadvantage of having a stable labour force, namely it is more difficult to change it to meet changing needs.
- Defective products can usually be rejected and replaced, with the only cost being that of the defective product itself, unless it has been allowed to pass on to the customer.

Application of quality to construction projects

Total quality management (TQM — see page 114) is a general approach to quality, and project quality is a specific case which requires particular interpretation. There are six areas of project work in which quality has to be examined.

(a) The quality of briefing and decisions by the client.
(b) Quality of the design process.
(c) Materials and component quality.
(d) Quality of project assembly.
(e) The quality of the project management activities.
(f) Project management as a means of promoting project quality.

This list is in general terms and is applicable to projects in any industry, but can be put into a construction context simply by changing the general phrase 'project assembly' in (d) above to the more specific 'building works'; all the other terms are applicable. Looking now at the six areas in turn, within the context of construction, consider first the items which involve the client directly.

The quality decisions

The client and the design team have an important role to play in the determination of quality on a construction project. This involves initially the fundamental decision on the standard that is set for the project as a whole, and it requires thought about whether the general standard is that of a sheik's palace or a site hut. More importantly exactly where in that spectrum does the project lie? There are smaller but still very significant differences between these extreme cases. Decisions on standards of building should be consistent, e.g. do not mix high-standard and low-standard materials — such as a softwood door in a hardwood frame, or a sophisticated software package capable of using colour with a monochrome monitor — remember the maxim that 'the strength of a chain is that of its weakest link'. The standard must be considered in relation to project life and the balance between capital and running or maintenance costs. It may well pay to use hardwood joinery which will need much less subsequent maintenance. One of the real difficulties here is that there is pressure on designers to keep building costs low, because 'someone else' is paying for the operation and maintenance, and they are not party to the design process. The whole subject of life-cycle costing is an important one which is discussed in Chapter 12. It considers the long-term sum of costs for design and construction, use and maintenance, and ultimate disposal. A major concern about decisions on standards is 'fitness for purpose', i.e. how well are the project and its components suited to the purpose for which they will be used?

Design quality

This is not so easily defined, and does not readily lend itself to sampling. Some of the points to note are listed below, but it is important to emphasise that the responsibility for the quality of the design process rests with the design team, whether this is an independent consultant, an in-house design team within the contractor's company or within the client's organisation.

- Actual design calculations can be checked for error or inefficiency, either by simple review, or ideally by a totally independent check.
- Component detailing is often regarded as a matter of 'good practice', and will frequently depend upon manufacturers' advice on applications and detail; e.g. the relatively simple example of laying, fixing, and flashing of roof tiles; there are very good documents provided by tile suppliers to show this — it is in their interest to ensure good quality on site.
- Quality of design is very important in terms of efficiency; are all materials and components compatible? (e.g. do not fix steel and aluminium together in a potentially wet location). Also, are materials and components of a reasonably consistent standard? — there is no point in building in any weak links.
- Quality attitudes can be encouraged in the design process; most architects, engineers and other designers are highly trained responsible people who can be educated in the philosophy of the total-quality approach.
- The responsibility for design quality lies with the design team, whether this is an independent consultant, in-house design group within the contractor's company or within the client's organisation.

Materials and components

These are the products of other industries, such as quarries, cement works, joinery manufacturers, heating plant manufacturers and so on; all of these use mass or batch production processes which lend themselves to the well-tried quality, planning and control procedures which have been established in manufacturing. Some of these supplier companies have taken on board the concept of total quality management, with the result that component quality has improved significantly, and can normally be relied upon when it is certain that effective quality management procedures are being followed. Some of the characteristics typical of batch or mass production which facilitate quality control are as follows.

- Sampling procedures are relevant, and with the analysis of trends can be used to predict and therefore prevent the manufacture of defective products.
- Acceptance sampling can be used if appropriate and required, with incoming materials and components inspected before they are incorporated into the project.
- Standards can be defined by the provision of a sample or prototype for inspection or testing, or by reference to the relevant standard, e.g. ISO 9000. (Note that in the past in Britain it was the British Standard which was used.) An example of prototype testing is that of non-standard window frames which usually have to be tested in a rig in which they are subjected to air pressure and water spray to examine for leaks; this is a special test which can only be carried out at a few recognised centres.

Quality on site

This presents a real challenge to the industry because of a number of factors.

(a) Most construction is one-off; it is seldom repeated exactly, and hence reproducible samples cannot easily be given and followed.

(b) Quality of construction is literally in the hands of site operatives, many of whom may not have been appropriately trained for the job. This means that it is necessary to impose fairly intensive supervision, and again supervisors are not always fully trained in quality matters.

(c) The use of labour-only sub-contractors (the 'lump' or 'grip') and simple bonus schemes concentrate attention on the speed of production and not on its quality.

(d) Variability of the production environment on site, e.g. weather, in itself militates against good quality and is a factor which brings about the need for 100% inspection.

(e) Mistakes and poor workmanship are not easily remedied — you cannot usually just discard defective construction as it has to be demolished and removed, then rebuilt.

(f) The time needed to demolish and replace defective work usually will lead to a delay in the project; often a case of one step forward and two back! Consequently pressure can be brought to bear to accept sub-standard quality in the interest of making progress; this should always be strongly resisted.

(g) Costs of defective construction are high, e.g. filling in an excavation dug in the wrong place. (If it is in rock then this would probably have to be with concrete.) Similarly the cost of breaking out reinforced concrete can be very much higher than the actual cost of replacement.

(h) Dimensional accuracy presents special problems, and should not simply be check-measured. (It is far too easy to fall into a trap when checking the position of a peg in the ground; an engineer may simply say to the chainman '17·450', the chainman thinks this looks like the reading on the tape but in reality it is 17·550, and the error is not big enough to be spotted by eye, and perhaps a whole line of columns is 100 mm out of position.) It is important that all setting out of site work is truly checked independently, e.g. by asking some other person to carry out a survey without having the drawing to consult.

Project quality management and quality project management

How do we manage the quality aspects of the project? It is helpful to start with a quality plan which sets out the standards to be achieved and the means of achieving them. Some of the methods are set out in later parts of this chapter. How well do we manage the project? Do we practise high-quality project management? Given that true project management can cover the whole of a project's life from 'concept to commissioning', it is the responsibility of the project manager to ensure that the quality of the management system is appropriate to the project. To do this, he must perform the following duties.

(a) Instill quality concepts in all who are concerned with the project, including the client.

(b) Set up quality procedures for purchases, i.e. materials and components.

(c) Undertake training programmes for designers, and carry out design audits.

(d) Provide training in quality procedures for site managers and other supervisors.

(e) Seek ways of relating payment to all aspects of performance, in particular quality as well as speed.

(f) Ensure that the work programme itself is of high quality, makes efficient and effective use of resources, and has been well thought out, questioned and discussed by all concerned. Having a good initial programme is only part of the approach; it is essential to ensure that as work on the project advances the programme is used to aid the control process and is itself kept up to date with the work. It is also valuable to carry out project audits from time to time, preferably at significant project stages rather than simply at arbitrary calendar intervals.

(g) Remember that time, cost and quality all interact. It is not possible to have the shortest construction time, lowest cost, and highest quality all at the same time, and decisions must be taken at the outset of the project.

Project management as a means of promoting project quality

One of the major obstacles to consistently high quality performance in construction is the supreme optimism of the people involved. This applies to both quality of work and to adherence to programme, and we often hear statements such as the following.

- 'We would have been up to programme if the weather had been good.'
- 'The project would have been on schedule if we had not encountered rock in the excavations.'
- 'It would have been possible to achieve a uniform appearance to the brickwork if we could have kept the same squad of bricklayers on the project.'
- 'A breakdown of the ready-mix concrete plant caused a long gap in supply, and this led to an unplanned construction joint in the floor slab of the water tank.'
- 'The bankruptcy of the cladding supplier led to a slight change of sheet part-way through the construction of the main hall — but you can only just see the join!'
- 'We know the wall is slightly off vertical, but it is quite safe and can only be seen from one direction — to knock it down would delay the project completion and there would be an argument about who was responsible for the cost.'

The above list of such examples could be extended by anyone who has worked in the industry. There is a great need for thinking ahead, with the main objective to get it right first time. This is where the concepts of project management can be brought into effect. Some of the basic principles of project management are

- think ahead of what is to be done
- consider what problems may arise
- seek ways of overcoming these problems
- examine the impact of deviations from the plan.

This process of thinking ahead is commonly applied to achieving progress on schedule, but it can equally be applied to the problems of quality management, and the rigour of project management thinking can be applied to the achievement of planned appropriate quality.

Thinking ahead means considering each aspect of each major component of the project, e.g. for brickwork is it certain that a full supply of consistent quality and colour of bricks is guaranteed? Is it necessary to buy the full quantity at the outset to ensure this? Can the same squad be retained for the whole of the work? Can some form of terminal bonus be agreed to keep the squad on the job to the end? Is the supply of mortar consistent? Consider what problems may arise. This applies particularly to areas of work where there are significant unknowns, e.g. foundations, drainage and other works below ground, or work which is particularly susceptible to bad weather, e.g. exterior decoration. In the first of these, lack of information may be a significant cause of problems, and can often be alleviated or overcome by better site investigation with more boreholes or trial pits. Consideration must be given to overcoming these problems. Some suggestions have already been made. Others may include having standby plans to deal with difficulties that might arise, and these can cover a whole range of provisions such as

- having surplus bricks, sheets of glass etc. to allow for damage
- preventing damage to stored materials by having proper protection
- having standby plant or equipment readily available, if not actually on site
- seeking alternative suppliers who can be called upon quickly, or perhaps using more than one supplier at a time
- using cranes with plenty of spare lifting capacity.

Impact of deviations from the plan   In spite of a great deal of thought along the lines above it is likely that deviations from the plan will occur — this is an almost inevitable consequence of the nature of one-off projects. Care should be taken to ensure that quality does not fall below the designed and acceptable standards in such cases, and that the consequential impact is on time or cost. There is a great risk that the anxiety to complete a project allows quality to slip. However, being late is a one-off fault which is unlikely to have later unforeseen effects; quality on the other hand is a continuing fault which remains for the life of the project. Quality standards once agreed should not be negotiable as an expedient to solve a time or cost problem. The practice of leaving quality problems to be sorted out later should never be accepted, and the once-common phrase 'we'll put it right later' is now unacceptable. The so-called maintenance and repair of buildings is a major activity of the construction industry, and represents a significant proportion of its workload. A high proportion of the work carried out in the maintenance period is in fact the correction of poor initial quality, either in terms of materials, components, or erection on site. General wear and tear can be accepted, poor quality must not.

The overwhelming need in construction is to use the methods of project management to ensure the following action is taken.

- Think of what may happen as well as what should happen.
- Avoid being a blind optimist.
- Make provision for deviations; quality is essential.
- Always seek to 'get it right first time'.
- Apply all the above to all stages; brief, design and build.

- Instill good project thinking and good quality concepts in the minds of all concerned.

TERMS IN USE IN CONSTRUCTION QUALITY MANAGEMENT

There are several phrases which are used within quality management. In the context of construction some of these are discussed below.

Inspection

'Inspection' is the examination or measurement of work during construction to check whether specified standards are being achieved. It does not of itself prevent or correct mistakes unless appropriate corrective action is taken subsequently.

Quality control

This is the step beyond inspection, and comprises the diagnosis of the causes of faults and their consequent elimination, the removal of defective work and/or the correcting of errors, and the anticipation of further problems. It therefore leads to a reduction in errors, but does not guarantee that all defects are eliminated. Quality control requires that good records are kept of the procedures, the test results and the actions taken together with information on subsequent tests and results. Both inspection and quality control may be thought of as curative, but it is well known that cures do not always work. What is really wanted is a means of preventing the occurrence of faults in the first place, namely a proper quality system. A comprehensive quality system in the context of construction depends not simply on inspection and control, but on a whole system which will include

(*a*) correct instructions to all, clearly communicated
(*b*) appropriate abilities and skills, hence training
(*c*) suitable, safe and effective equipment
(*d*) good site working conditions, with proper protection
(*e*) checks or tests on completed work, properly recorded
(*f*) the power and authority to correct faults
(*g*) motivation to produce quality
(*h*) a document system that records pass/fail
(*i*) confirmation that faults have been remedied.

Quality assurance

This is a phrase widely used to mean the overall range of activities which are concerned with achieving quality. (Note however the more specific acronym QA which has a particular meaning in ISO 9000 and is discussed later in this chapter.) Quality assurance includes all of the above-listed functions, but in addition is aimed at establishing the right environment within the whole organisation, involving everyone in it. This ranges from the members of staff who should answer calls politely and promptly, through all the initial discussion of a project, into design, detailing and construction. It includes efficient handling of goods in and out of stores, plant maintenance, wages and bookkeeping; every activity of the enterprise. It is a total commitment to quality that will lead to the construction of a quality building which meets all the requirements of the client and user. It is a step towards total quality management, which is a commitment to continuous improvement in all the work of the organisation.

Total quality management (TQM)

In recent years there has been a very significant move to improve the quality of a wide range of products (Conti, 1993). Much of the emphasis for this has come from the international industries such as motor vehicle production and computers. International competition has focused more and more on the reliability and performance of products rather than simply on their general features and price. Japan has led the way in developing new approaches and attitudes to the management of quality, and much of the work started in Japan has now been adopted in other industrialised countries. Most of this development has taken place in manufacturing and process industries, but some of it has been adopted within construction and other project-based industries. As was noted earlier in this chapter, the construction industry is a consumer of many materials and components produced in manufacturing and process work and therefore becomes involved with appropriate quality procedures. Many of the detailed techniques are hence of interest to project management, and are discussed later in this chapter. First it is relevant to consider briefly what is meant by TQM. It is sometimes simply stated as a total commitment by an organisation to the continuous improvement of the quality of its products and/ or services. This is not confined to the quality of its actual products but is extended to everything that the organisation does. This approach has very wide implications, as it includes such things as the way in which the telephone is answered; this is indeed important to a company as the telephone is often the first point of contact with a company and can start relationships either well or badly. This is also true of written communication which should be prompt and clear and have a good general appearance; a company's office should be clean, tidy, welcoming and efficient; its staff relations and training need to be good to achieve all this. In short the view is taken that only a high-quality organisation in a high-quality environment can be relied upon to consistently produce high quality products. During the 1960s a study was undertaken by the Building Research Station (now the Building Research Establishment) of a number of major buildings, mostly in the London area, and including the headquarters of the Institution of Civil Engineers and of the Royal Institute of British Architects. Among much other useful comment it was stated that '... most inspired and durable works ... were not created by such tools (i.e. books of rules etc.), but by the intuition combined with keen observation of past performance and commonsense ...'. This statement is entirely consistent with the general objective of TQM, and illustrates that the existence of a complex quality assurance scheme will not alone give a guarantee of good quality.

TQM is not however simply a matter of installing good equipment and people in a nice building; it is important to generate the ethic of high quality throughout the organisation. It is also necessary to have a detailed set of procedures to ensure that everything that is done is in accordance with the appropriate standards. It is fair to say that an overall positive approach to quality and an effective Quality Assurance system are both necessary to achieve high quality in a project; each one on its own is necessary but not sufficient.

The process of Quality Assurance is now highly regulated through the ISO 9000 standard, and an organisation can seek to qualify through this body to demonstrate that its work is carried out within strict quality procedures. The phrase 'Quality Assurance' is often used to describe this approach, but this should not be confused with the rather less specific phrase quality assurance used earlier in this chapter (without capital initial letters) which has for a long time simply meant the application of control to an individual process. ISO 9000 has become very important to many companies in the manufacturing and process industries, since it gives an implied guarantee of the quality of a company's product (but not an explicit guarantee). Because of this many purchasers of goods or services will only buy from companies having ISO 9000 registration, and it has therefore become essential for suppliers to get this accreditation in order to maintain their market. Many such suppliers, especially small ones, find it an onerous task, and there has been much concern about the time and cost involved in conforming. The standards apply not only to manufacturers but also to design firms, many of which have now become registered. The position in construction and other projects is however less clear. Turner in *The Handbook of Project-Based Management* poses the question of whether in one-off work it is justified to make great efforts to prevent any defective work, as compared with repetitive production where the benefits will be gained over a long run. There is a considerable dilemma here, which in turn is reflected in the extent to which formal standards have been taken up in the construction industry. Professor P. Barrett of Salford University has stated that a survey showed that when clients are choosing a builder only 3% will take formal standards into account, and the great majority will not expect to pay more for work by a registered contractor. ISO 9000 is a somewhat complex set of documents, and as was stated earlier it had its origins in manufacturing industry. Relatively little has been published on the application of international standards to construction, but a very useful book, *A Guide Through Construction Quality Standards* (Atkinson 1987), was published in 1987. This makes clear some of the ways in which quality in construction is different from that in other industries, and presents a challenge to project-based work. In addition to the discussion of quality this book provides much useful information on the various standards in use not only in the UK and Europe, but also in North America and Australia.

The process of becoming registered for ISO 9000 is in three stages. First a company must examine carefully the way it works and what procedures are involved to plan for, implement, and check quality. Second it is necessary to document all the procedures and the ways in which they are operated, then third an outside agency has to examine and approve the work done in the first two stages. Further

115

information on the operation and implications of registration can be found in several publications, especially those of the British Standrds Institution (see bibliography).

THE COST OF QUALITY

The two major aspects of quality set out earlier have to be considered separately. Firstly the decisions on standards to be specified will clearly have cost implications; for example the prices of different standards of materials vary widely, and there will be a wide range of projects from the most palatial to the most humble. This subject of the interaction of quality and cost refers not only to the actual materials used but also the performance of the finished project in relation to its use, maintenance and life. This is a distinct aspect and relates largely to the initial decisions on quality at the briefing and design stages, and it is discussed more fully in Chapter 3. At this point consideration is restricted to the costs associated with the achievement of the standards once they have been set, i.e. the implementation of the quality management system. It is this latter area that is often referred to as the cost of quality, and can be divided into three main areas. It is generally the case that the sum of these three costs should be minimised, rather than seeking to reduce each of them separately. The three areas are the cost of failure, the cost of control and the cost of prevention.

The cost of failure

When faults arise in a project there will be a range of costs to be covered, as was indicated earlier in this chapter. A number of studies of construction faults have been undertaken and these show interesting results. An HMSO publication found that costs of poor quality in building amounted to 10–18% of total cost in an industry where profit is less than 1·5%. On one specific project the cost of poor quality was put at 11%, and this could have been substantially reduced by preventive expenditure of 2·5%. Another study showed that 50% of faults originate at the design stage, 40% relate to site work, and less than 10% to materials quality. 25% of all faults were due to poor information provided to the site, and less than 10% were due to new materials or methods. Further useful reports are given in a reference in the bibliography (European Foundation for the Improvement of Living and Working Conditions, 1991).

There must be many examples of errors which can be quoted by any reader of this book. Rather than give a long list a few typical examples are given, where faults either turned out to be very expensive, or had the potential to do so.

In the case of a high-rise flat development the building was clad with pre-cast concrete panels which were fixed to the structure with bolts. The material of these was not properly specified and the contractor used mild steel bolts coated in bitumen. There was no problem for many years, but with time the panel joints began to leak water, rusting commenced and eventually a panel fell off. The risk to the public was very great, and the building owner was faced with initial expensive emergency measures and then with re-fixing the entire cladding system. This was a fault at the design stage. In the design of the fire-pumps on an off-shore oil platform the pumps were suspended just below the surface of the sea, so that they were permanently submerged, and were to be fixed using special non-corroding bolts made of an expensive alloy. During construction someone substituted mild steel bolts for the alloy ones and this was not noticed. It was only when a test was set up for the fire system that it was discovered that one pump was missing, the bolts having corroded, and the pump had fallen into the sea. Fortunately the problem was found at the expense of one lost pump and the re-fitting of all the others; it could have been much worse if a large fire had broken out on the platform.

The costs which have to be considered regarding faults in construction include the following.

- Repair in cases where this is possible, and replacement with new work in cases where repair is not possible. Anyone with experience of construction projects will know that mistakes in such cases have some additional costs and may be caused not only by actual faulty site work, but by mistakes in briefing, design, detailing and communication, i.e. any preceding stages of the project.
- For building work actual demolition may be necessary. Reconstruction following demolition can also interfere with adjacent new work, and lead to problems of damage, dust, access, and making good the join between old and new work.
- Delay in progress of the project may be one of the less obvious costs but is certainly a real one. The disruption of programme could impact in relation to the availability and use of resources.

There is no doubt that in construction the consequence of errors made at an early stage may not become apparent until much later, for example errors at the briefing stage may not be realised until commissioning at the handover stage, and detailed design may be completed long before construction. It has been suggested that half of all errors occur during the initial briefing and feasibility stages, and a further one third in the run-up to the start on site. This implies that only about one sixth of all errors are initiated on site, but it is there that most consequential costs occur; the site is the place where all the earlier mistakes come together. There is often a temptation to look for improved construction performance on the site only, whereas design and procurement changes would sometimes be more appropriate. It is of interest to note that a Building Research Establishment survey indicated that the most important quality decision on a project lies with the client in choosing a designer; this could be extended to say that the decision on procurement route is the means by which this is achieved (see Chapter 11). In this regard project management has a service to offer to clients, in helping them choose not only the most suitable designer, but also the most appropriate contract structure in relation to both design and construction.

The cost of quality control

This is the sum total of inspection costs including tests and the reporting thereon, the delays caused while work is inspected, and the overhead costs of maintaining and operating a quality control system. There is no doubt that it is possible to spend excessive time and effort on control procedures, and traditionally it was normal practice to calculate the level of control at which the sum of control costs and failure costs was at a minimum; beyond this point it was felt that the value of detecting more faults was more than offset by the increasing cost of detecting them. This approach was more relevant in the manufacture of simple products where the costs of rejecting defectives was low, and the consequences of allowing defectives to remain undetected were not serious. For example, if a machine making woodscrews occasionally produced a screw without a slot in its head, the result is that the joiner would simply throw it away. It would be impossible for such a defect to be incorporated into another product, and problems would only arise if the number of such defects became a real nuisance to the joiner. By contrast the processes for the manufacture of a vaccine would have to be very tightly controlled because of the potentially severe consequences of allowing defective goods to be used. In such cases no remedial action would be possible, and cases of permanent injury or death could be followed by claims for the payment of large compensation claims. In most projects the magnitude of consequences would lie between these two extremes, but would cover a wide range.

The cost of prevention

As is described earlier, to totally prevent the production of defective work is the aim of TQM, which is a complex and costly way of working. What has to be decided in any project organisation is whether or not this cost can be justified. The development of TQM has mostly taken place in the multi-national manufacturers, especially in Japan. It has been felt necessary for them to operate at a high quality level, partly because of reputation in the market and seeking a quality edge on competitors, but also because the high-volume output permits the spreading of the costs of the quality system. It is clear that the same approach to quality would be relevant in the case of projects in the computer industry and it is indeed the case that quality systems are very advanced there. It has been stated that if a mistake is made at the design stage in a computer installation and is allowed to proceed to manufacture then the cost of correction is increased by two or three orders of magnitude. If it goes further into the final installation and use then there may be a further thousand-fold cost escalation. In such situations the importance of the prevention of faults can be seen to be vital.

The case of construction is less clear, and much work has still to be done to establish the appropriate level of quality management. It is already clear that the TQM approach as used in manufacturing is not fully appropriate in construction, due to the one-off nature of much of its work. It is quite possible that the cost of preventing a defect in the construction of a wall could exceed by many times the actual basic cost of building it. To take an extreme approach it might be thought necessary to determine accurately the position of every brick laid, either by actual measurement or by the use of a jig. This would in itself cost a great deal and would slow down the building of the wall. It is certainly easier and quicker to rely on the eye and dexterity of the bricklayer, and given that good training and supervision is provided, together with accurate and clear instructions, then good quality work should be completed. There remains however the problems already referred to of the costs of remedial work in construction, and the present difficulty lies in determining where the correct balance lies.

A further problem of quality in construction derives from the fact that defective work may not come to light until a long time after the actual work was done. Errors in underground services may literally lie buried for years before a new building is occupied and the user finds faults. It is perhaps then difficult and expensive to correct the fault, especially if the defects liability period has expired. Some recent cases have placed liability on designers and contractors over very long periods, but this threat of long-term liability is not the best way of ensuring good work in the first place. It has been suggested by Turner (1993) that one way round this is 'for the prevention costs to be borne by the parent organisation as an overhead, with the whole organisation benefiting as savings feed back into more effective projects'. If this means that clients will have to pay the direct costs of preventive measures it will take a lot to persuade them of the correctness of this approach.

QUALITY CONTROL TECHNIQUES

Earlier in this chapter reference was made to the quality control techniques which are used in manufacturing and process industries, and which are therefore relevant to the production of materials and components for the project-based industries. These include some materials which are produced on the site of projects, and it is therefore useful to look at some of these procedures to see where they can be of use in project management.

Sampling

One of the most common methods of examining materials is the procedure of testing samples of materials to ensure that they comply with the specified standards. Some of the properties to be examined, e.g. ultimate strength, will involve destructive tests, and clearly in these cases the testing of samples provides the only practicable method. Even where tests are non-destructive it may prove to be the only economical way to assess the properties of the material or component. Consider as an example the testing of facing bricks to ascertain whether they are in accordance with specification. Some of the properties to be tested would include

- dimensional accuracy
- density
- water absorption
- ultimate strength
- colour
- surface texture.

Standard procedures exist for the selection of samples, and for the size of samples. There is an important point here that should be mentioned; the confidence that can be placed in a sampling process depends upon the size of the actual sample, and not on the proportion that the sample is of the whole batch being considered. In the jargon of statistics the whole batch would be referred to as the 'population'. It is often tempting to think in terms of testing a fixed proportion, say 1%, of the whole batch, but this is not the correct measure. A sample of 10 bricks tested to destruction will give the same measure of the whole batch, irrespective of whether the batch contains 1000 or 10 000 bricks. It is important when specifying sample sizes that this is remembered.

100% testing

This is an approach used in some processes, but clearly is not possible where the test is destructive. It is however used in cases where the test is simple but important, for example the automatic measurement of a component being produced by a machine. It may also be used where sampling is not relevant, for example in unique tasks in a project where a sample of only one is all that exists, and it amounts to 100% in any case. The alternative in such situations is to do no testing at all for many of the tasks, and rely on the general skills in use and the fact that other similar (but not identical) tasks are being tested. This is in effect what happens in much of construction, and it is not really satisfactory in many cases. Many specifications call for all work to be inspected, without stating what that inspection actually entails. All too often it amounts only to a cursory visual inspection which is not quantified or properly recorded, and may consequently be inadequate. This is again part of the challenge to the construction industry to improve both its quality and its quality assurance.

Test results

Results can be used in a number of ways, and to understand these fully it is necessary to be familiar with some basic statistics; there are many good texts on this subject (Hamburg, 1974). It is usual to work out from tests the properties being examined, strength, weight, density etc. (these are referred to as the 'attributes' of the product being tested) and then calculate a 'mean' (i.e. average) value for each sample. The size of the sample will give a measure of the confidence that can be placed in these values, obviously if the sample is very small there is a possibility that an occasional extreme value for the parameter will get undue emphasis. With a large sample there is a good chance of a range of values being included with the result that the average is a fair representation of the whole batch.

What is sometimes of more use than the actual values of each parameter is the way in which values of sequential samples change. This permits an 'analysis of trends' which can often be used to provide early warning that a particular parameter value is likely to break through its permitted limit in the near future. In this way it is possible to take action before any defective items are produced, and hence take truly preventive action. The way of doing this is set out in Fig. 38, in which the following terms are used.

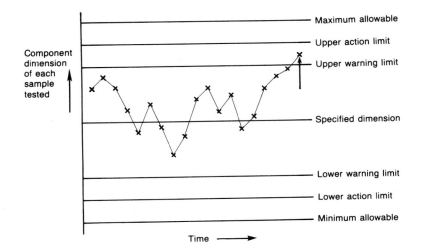

Fig. 38. Quality control chart for sequential samples of a component part

'Upper warning limit' is the value of the attribute being measured which, while still acceptable, is the value at which the process seems to be drifting. A closer look is needed, perhaps by a bigger sample or a more rigorous test.

'Upper action limit' is the value at which the process is liable to make defective product and hence action is essential to reverse the trend. This is often achieved by stopping the process until the cause of the problem is found and eliminated.

On the quality control chart of Fig. 38 a record is shown of one dimension of sequential samples of a component which has been tested. At the point indicated by an arrow the upper warning limit has been crossed, and further investigation is needed. The trend of the previous four samples tested has been steadily upwards, implying that something in the production process has started to go wrong and must be corrected. If this is done quickly it will prevent the manufacture of defective items.

'Lower warning limit and lower action limit' are the corresponding values in the other direction where there is concern about attribute values below that which is acceptable. In some cases there may only be concern about values which are too low, in which case upper limits are not used, but even where the attribute is something like strength it is useful to know when it is drifting higher than required because it implies that the process is out of control.

'Sample range.' Another useful measure is the range of values found within a sample, i.e. the difference between the highest and lowest values. Again this should not be excessive as it indicates the extent to which the process is in control and hence the confidence that can be placed in the mean values of the sample.

Further exemplification of standard quality control methods can be found in many books (Woodward, 1982; Hamburg, 1974; Johnson et al., 1974; Hall, 1983).

POSTSCRIPT ON QUALITY IN PROJECTS

The two following quotations sum up much of the thinking behind quality management in project work.

'If all things were done twice, all would be wise.'

'We do not seem to have time to plan the work properly, but we seem to have time to do it twice.'

8 The interaction of time, cost and quality

In the previous three chapters we have considered respectively the topics of time, cost and quality; it is well appreciated that these three aspects do however interact, and this interaction forms the subject of this next chapter. Consideration is given to the general ways in which they impact upon each other, and then it is shown how calculations can be made to help with the management of the interactions and advantage taken of them.

TIME AND COST

Many project managers will be familiar with the wish that clients express to have the highest possible quality of work carried out as soon as possible and at the minimum cost. It does not take a great deal of thought to realise that these three objectives are not simultaneously compatible, and that compromises must be made in some form of trade-off. For example if a project is really needed at the earliest possible date then it would be expected that it will be operated on a 24-hour day schedule, with more than enough manpower available, and stand-by equipment on hand to cover for any plant breakdowns. This can be done but will entail additional costs of overtime work, and surplus labour and equipment.

QUALITY AND COST

Another familiar situation arises during the briefing and design stages of a project where there may be a conflict between the performance of a component and its cost. For example when specifying a machine to carry out a task, whether it be a microcomputer, a milling machine in an engineering works, or a mixing vessel in a bakery, there will be a range of machines available. There may be many factors to be considered in making the choice, but among them will be performance and reliability compared with price. Chapter 7 dealt at some length with the importance of the initial standard to be specified at the design stage, and it is this aspect of quality that enters the debate about which type of machine to specify, and hence it is clear that there is an interaction between quality and cost.

QUALITY AND TIME

A third interaction is possible, namely that between quality and time, but this usually manifests itself in a rather unfortunate and generally unacceptable way. Imagine a situation on a project where a section of work is being done and is taking longer than had been planned, for example the writing of some software, or the construction of the roof of a building. In each of these cases there is pressure brought to complete the section of work as quickly as possible as it is on the critical path and is holding the project back. While work is very nearly complete it really does require more time to ensure that it is up to the specified standard (free of bugs in the case of the software and watertight in the case of the roof). In this situation we are dealing with the second aspect of quality which was discussed in Chapter 7, namely the achievement of the standard or performance required. In the two cases cited here there is a need to spend more time to meet the required standard, and hence there is an interaction between time and quality. There have been a few famous cases of computer applications where the pressure to go on-line has forced implementation to go ahead before full testing has eliminated all the potential problems.

THREE-WAY INTERACTION

It has thus been shown that there is a three-way interaction between time, cost and quality and this is sometimes represented by the triangle shown in Fig. 39 which shows that for any one project there will be a balance struck between the three factors. This can be represented by a point within the triangle, not drawn to any scale but simply indicating the relative importance given to the three factors. The diagram can be interpreted in the following way, with the relevance of the points marked on the figure

- at Q there is total dominance of quality
- at T there is total dominance of time
- at C there is total dominance of cost
- at A all three factors have equal weight

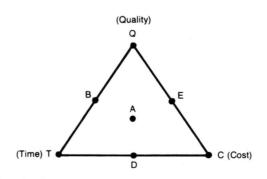

Fig. 39. Time/cost/quality interaction

- at B cost is of little concern, time and quality dominate
- at E time is of little concern, cost and quality dominate
- at D quality is of little concern, time and cost dominate.

In practice most projects will be placed somewhere near the point A, because usually a reasonable standard is required, costs must be affordable and projects can rarely be allowed to run on for very long. Some illustrations may help to clarify the way in which different projects are given different emphasis.

In the development of a new vaccine for widespread public use it is clear that quality must be the predominant criterion, with such a project being placed at or near Q, with a willingness by the client to continue allocating adequate resources and to allow work to run for as long as is necessary to give confidence in the quality of the output. By contrast in the case of a natural disaster such as flood or earthquake there will be an initial total emphasis at the rescue stage on time, with little regard for cost or quality, indicated by point T in Fig. 39. In another field an important public building such as a millennium-funded gallery or museum would perhaps be indicated somewhere close to the line between Q and B, showing considerable emphasis on quality, some concern about time, but a willingness to pay the relevant cost. This does not mean that costs are not carefully controlled once they have been agreed, simply that the design of the project can be carried out knowing that money is available to do a very good job.

Making use of the interdependence of time, cost and quality

It is important that the interactions are understood, and that at the start of any project a deliberate decision is taken on where in the triangle of Fig. 39 the project should be placed. This decision does raise something of a problem for project managers however, since it is necessary to focus on whether the interactions are simply recognised or whether they can be deliberately used in project management. There is no doubt that the three factors do naturally interact, but some practitioners have questioned this view. The argument they put forward is related to quality, and in particular to the achievement of the standards set at the briefing stage. The case they make runs roughly as follows: production should always make every effort to achieve the required standards at all times, and that to try to save money by cutting corners on quality just does not work. This is because the consequences of falling below specified standards will usually incur additional costs in terms of re-work, delays, and perhaps even direct penalties. Hence it is argued that quality cannot be traded off against cost or time savings, an argument which is entirely consistent with what is being said in this book. Here the case is made that it is at the design and briefing stage that decisions have to be made which do take account of the fact that a high standard of specification will often involve more time, and nearly always will cost more money. It was clearly stated in Chapter 7 that once decisions on standards have been made and the project work put in hand, then those standards are no longer negotiable and must be achieved irrespective of time and cost. It is important to remember that poor levels of performance achievement will have repercussions throughout the working life of the project installation, but in most cases an overrun of time or cost is a one-off happening from which recovery is possible.

The purpose of the above paragraph therefore is to agree that quality cannot be traded off once the project installation is under way, but also to state that such trade-offs are feasible at the briefing and design stage. This shows only partial agreement with those writers on project management who have decided to take the quality dimension out of the triangle. There is of course a rather more extreme view which can be taken that any project of major significance should always be completed at the highest possible standard in every way, irrespective of the cost or time taken. Such a view may well have been taken by the builders of some of our ancient monuments, but it is not one which would gain much support in our modern commercial and highly competitive world. It is a view however which does have some attraction for purists, and may be worthy of our attention, especially where consideration is given to the environment and conservation of resources.

CALCULATIONS BASED ON TIME/COST INTERACTION

It is possible to obtain some useful results by analysing the ways in which time and cost interact in a project, in particular to seek an optimum duration for a project. First it is necessary to introduce some specific terms which will be used. The 'normal duration' of an activity is the elapsed time required to complete that activity if it is approached in a 'normal' way, that is to say without any pressure to complete it quickly, using a balanced team working standard hours, but without any unusual delay. Associated with this normal duration will be a 'normal cost', shown as point A on the diagram of Fig. 40; this cost is usually based on the direct costs of labour, plant and materials. For the purpose of the immediate discussion overhead costs are excluded, but will be covered later in this section. It is logical to imagine that the duration of an activity could be reduced at some expense by any one of a number of means, possibly by working a few hours overtime each evening and bringing about a reduction in duration from 6 days to 5 days. This will have a small extra cost because of the premium pay rates for overtime, and this is shown as point B in the diagram. If

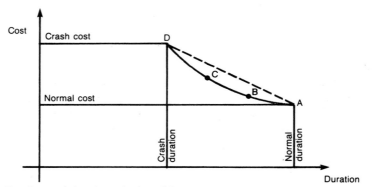

Fig. 40. Time/cost relation for a single activity

a reduction of the duration by a further day to 4 days is required this might be achieved by increasing the size of the work team. It is likely that the team would work more quickly but less efficiently because it is no longer balanced, and would therefore raise the cost a little more steeply, as shown by point C in the diagram. In the event that there is still pressure to carry out the work even more quickly it could perhaps be reduced by one more day by doing some of the work using a nightshift at a rather higher cost, as at point D. This might then be considered to be the absolute minimum duration for the activity and it is then called the 'crash duration' which is the minimum feasible time for the activity; it has associated with it a 'crash cost'. By this means it is possible to draw a curve representing the relationship between time and cost for each activity in the project.

The practical context

It is necessary to put the above mathematical curve into a practical context. First, a mathematical curve should perhaps extend to values to the left of point D, but we have defined D as the minimum possible duration. Second, it is clearly possible that the duration could extend to the right of point A, but if it did so then it implies an 'abnormal' duration and would probably raise costs, albeit slowly. Point A, that of normal duration, should surely have been found by doing the work in the most economical way, and it would never be desirable to deliberately extend the activity duration beyond that time. Another practical aspect is that while each day's reduction in duration will imply an increasingly greater cost, it is not realistic to give this a specific mathematical form. There may in some cases be no intermediate points between A and D, with only either a slow normal way or a quick crash way of doing the work; for example in the excavation of a hole it might take a man with a shovel 1 day or a mechanical excavator 1 hour, with no meaningful intermediate duration. For these reasons and in order to simplify the mathematics it is usual to draw a straight line between A and D, (broken line in Fig. 40), and then it is possible to represent the relationship between duration and cost by the following expression. The additional cost of reducing the duration by one time unit

$$= \frac{\text{crash cost} - \text{normal cost}}{\text{normal duration} - \text{crash duration}}$$

Note that it is clearly important that all costs on a project should be defined in a consistent way, but this does not need to involve totally comprehensive costs. For example, in some cases of looking at the time/cost interaction it may be possible to exclude all materials costs since these will be independent of the duration of the activity; there will always be the same amount of concrete in a factory floor, irrespective of the time taken to pour it. In many projects it may be adequate to consider only equipment, labour and management costs since these are the ones most likely to change with acceleration. This is not universally true, however, if acceleration is to be achieved by incorporating a different piece of equipment or software; for example buying an item 'off the shelf' may be more expensive but quicker than creating one 'made to measure'.

Worked example of project duration optimisation

The tedious arithmetic and the limited final value may make this example seem rather daunting. Readers may rest assured, however, that the calculations are very elementary, and that in practice they could be carried out by computer. They may not even be necessary, but the method does give a very good feel for the way in which project durations can be reduced in the most economical way. Aspiring project managers are therefore asked to accept that it presents a good way of understanding what can be done, and a rather more pragmatic approach will be described later in the chapter. The simple network in Fig. 41 has durations and costs as set out in Table 7. The values in the last column are calculated using the above expression (page 127) and the values for normal and crash durations and costs are in each row.

The following calculation now proceeds step by step, seeking at each stage to reduce the project duration by 1 day in the cheapest way. The first step is to calculate the earliest and latest event times for the network on the basis of 'normal' durations, using the methods described in Chapter 5, with the results marked on Fig. 41. Note that the critical path passes through event numbers 1, 2, 4, 5, 6 and also that beside each activity is the calculated 'cost per day reduction' for that activity. It is already known that in order to reduce the total duration of a project it is necessary to reduce the length of the critical path, and we now seek to do that in the cheapest possible way. Inspection of the cost per day reduction values along the critical path will show that the lowest value is £5 in the case of activity 4–5, and therefore this is the activity where a reduction in the project duration can be achieved by the expenditure of £5. Having done this it is then necessary to re-calculate event times and to check the route of the critical path.

If we set out to find a second day's reduction it is tempting to reduce the duration of activity 4–5 by another day, but we can see from Table 7 that it has already been reduced to its crash duration and no further reduction can be made. It is now necessary to find the next cheapest critical activity. This is 2–4, which has a cost per

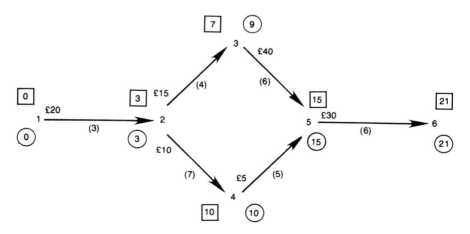

Fig. 41. Network for time/cost example, with normal durations

Table 7. Duration and costs for network in Fig. 41

Activity	Normal cost: £	Crash cost: £	Normal duration: days	Crash duration: days	Cost per day reduction: £
1–2	£5	£25	3	2	£20
2–3	£75	£105	4	2	£15
2–4	£55	£85	7	4	£10
3–5	£120	£200	6	4	£40
4–5	£30	£35	5	4	£5
5–6	£140	£200	6	4	£30

day reduction of £10. Again the event times are re-calculated and the critical path checked. It can be seen from Fig. 42 that a second critical path has been created in parallel with the previous one, and that all activities are now critical. Hence it is not possible to reduce the project duration simply by taking a further day out of activity 2–4 as the project duration will still be kept at 19 days by the path 2, 3, 5, which is now also critical. It would involve taking a day out of this path as well as out of 2–4 to bring about a reduction of 1 day and this can be done most cheaply for an additional cost of £10 + £15 = £25. However, if the rest of the activities are examined it can be seen that a day could be taken out of the critical path more cheaply by reducing the duration of activity 1–2 at a cost of £20. It should be noted in doing this that 1–2 is now at its crash duration of 2 days and cannot be further reduced.

The same procedure should be continued until no further reduction in project duration can be achieved. It would seem unnecessary to describe this in full detail, but some readers may like to follow this logic through. The final network and event times are shown in Fig. 43, and Table 8 shows the values of costs for successive reductions in project duration.

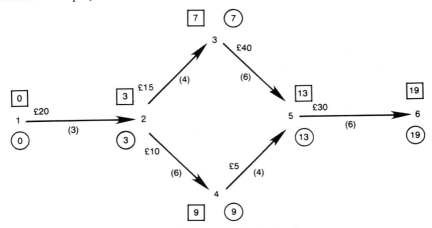

Fig. 42. Network for time/cost example after two days' reduction

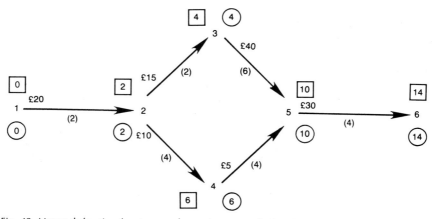

Fig. 43. Network for time/cost example, project at crash duration

Table 8. Values of costs for successive reductions in project duration

Day saved	Activities reduced	Additional cost: £	Note
1	Reduce 4–5 by 1 day	£5	4–5 now crashed
2	Reduce 2–4 by 1 day	£10	path 3, 3, 4 now critical
3	Reduce 1–2 by 1 day	£20	1–2 now crashed
4	Reduce 2–3 by 1 day and 2–4 by 1 day	£25	
5	Reduce 2–3 by 1 day and 2–4 by 1 day	£25	2–3 and 2–4 now crashed
6	Reduce 5–6 by 1 day	£30	
7	Reduce 5–6 by 1 day	£30	5–6 now crashed and a crash path through project
Total reduction of 7 days		£145 total cost	

Significant points to be drawn from Table 8

When a real project is under extreme pressure to complete as quickly as possible there is a temptation to crash everything, and it is possible that in the rush a critical activity does not get the resources it needs. In this example crashing has been applied only where it gives results, and in particular note that activity 3–5 has not been accelerated at all, and it had the highest 'day saved' cost. Selective acceleration is preferable to the panic reaction of 'crash everything' which has sometimes been used with expensive results.

The maximum reduction in time of this project is 7 days, giving an overall project duration of 14 days, but any intermediate reduction may make sense. For the sake of illustration assume that the project overheads amount to £22 per day, and now consider whether it would make sense to reduce the project to 14 days duration. The total additional cost of the 7-day reduction has been shown to be £145 (see Table 8), and this would result in an overhead saving of 7 days at £22 per day equalling £154. Hence the net overall saving is £9 (£154 − £145). While in absolute terms this may seem trivial it is of value relative to the cost figures in this exercise. There does remain the question however of whether this is the maximum saving and optimum duration of the project. Inspection of the table will show that the first day's reduction can be obtained at a cost of £5, and with an overhead reduction of £22 the net saving is £17. Each of the first 3 days' reduction will show a net benefit, but the fourth day's reduction will cost £25 but only save £22, a net loss of £3. The method of finding the optimum then is to scan down Table 8 (which is set out in order of increasing 'cost per day saved') until that cost exceeds the daily overhead rate. Following this procedure in the numerical example would indicate an optimum project duration of 18 days, and a total project cost made up as follows.

'Normal' project cost	£425
Extra cost of reducing by 3 days, £5 + £10 + £20 =	£35
Overhead cost, 18 days at £22	£396
Optimum project cost	£856

Finding the optimum project duration may be of interest at the design and planning stage when there is only an outline or master network which does not have a large number of activities. The procedure outlined above does generate a great deal of work, not least because it needs four data items per activity, namely the normal and crash durations and costs. The calculations are lengthy and even where a computer is used the data input is significant. A rather more common use of the time/cost interaction is in a case where a project has to be accelerated and it is desired to find the most economical way of doing this.

Applying the time/cost concept without detailed calculations

Earlier in this chapter it was stated that an understanding of the time/cost trade-off is of value and does not necessarily involve the lengthy calculations set out in the above paragraphs. A rather more selective approach will first of all recognise that only activities on or near the critical path need to be considered, and in most practical networks this means only about 10% of the activities will require detailed cost calculations. Furthermore it will be obvious that some activities cannot be accelerated, for example statutory delay periods, and can be ignored. The result of this narrowing down can then turn into an approach which is one used in operational research, and is as follows. Given a sum of money, say £100, what is the most effective way of using it to reduce the project duration? An experienced project manager will be able to scan along the critical path and select candidate activities for acceleration, possibly picking those with a reasonable duration and a relatively low plant and labour cost. There is clearly not much point in selecting short activities since they have little reduction potential; those with high labour and plant costs are likely to be expensive to accelerate.

Time/cost interaction in practice — a case study

A small manufacturing company occupied a number of buildings on a fairly congested site, but with some limited room for expansion into new buildings. At the time that this project arose they were also occupying a production shop rented from a neighbour on a short-term lease. They were well aware that at some time this lease could be terminated and had therefore taken the precaution of designing and obtaining planning permission etc. for a new building on their own site. Finance for the project gave them some concern and the rent they were paying was fairly low so they decided to shelve the project until it became essential; there were therefore no tender documents. The company had always expected to get plenty of notice to vacate their rented premises, and anticipated no problems in putting up the new building when it was needed. Somewhat to their surprise the building owners gave them three months' notice to terminate the lease, as the building was urgently needed for their own purposes. It was then decided that the new building should be erected as soon as possible, with the target of a construction completion within 11 weeks to allow 2 weeks to move equipment and resume production.

The company's consultants recognised the need to take a 'fast-track' approach and to short-cut normal procedures, and recommended that contractors be approached immediately to negotiate a contract based on a fixed price and a construction period of 11 weeks. As part of the negotiation it was stated that a penalty for late completion would be imposed, but the contractors were not happy with this and only agreed if a substantial bonus for early completion was offered at the same time. The client felt safe with this proposal since it was unlikely that a time much below 11 weeks could be achieved, and an extra week or two for moving would be valuable. The contractor that was selected looked carefully at the project, which consisted of a simple shed with a pre-fabricated steel frame and metal sheeting, and concluded that on the basis of 7-day working it would be possible to complete the construction to the stage of allowing plant installation to commence in a period of 9 weeks. The contractor submitted a price which was accepted on the basis of the times and penalty/bonus stated, and work was put in hand immediately.

A network of the project was drawn, and it was no surprise to find that erection of the steel frame was on the critical path. What was a surprise however was that despite the existence of full working drawings the delivery of the steel frame was quoted as 6 weeks, which was clearly going to delay the project. The contractor immediately visited the steelwork fabricator and pointed out the urgency, but the fabricator simply responded that all their projects were urgent and that this was a very small contract for them and they were not prepared to let it jump the queue of work. In the ensuing discussion the contractor enquired about the use of overtime working, but the supplier said it was not their usual practice and their prices did not allow for it. From further discussion emerged the proposal that this small order would be taken out of the normal work schedule and worked on at weekends only, with the contractor paying the additional cost of doing so. This meant that the existing fabrication workshop schedule was not changed in any way, but the small fabrication job was completed over the next three weekends and was available on site when needed. The project was completed within the planned 9 weeks, the client was well satisfied and the contractor was paid a bonus. The clue to this success was in identifying where the spending of a reasonable sum of money would have a significant impact on reducing the project duration.

The principle described in this case can be applied in many situations without undertaking lengthy calculations. Other examples are listed below.

- In many installations of new equipment a point arises where an existing power supply has to be cut off and everything connected to a new source. While this work is being done little else can be achieved, and the project is in effect suspended until this task is complete. Usually only a few people will be needed, and therefore it pays to provide as much support as possible to ensure speedy work; it is also common practice for this sort of work to be carried out overnight so that there is little or no delay to the project. In this way a small extra sum is paid to reduce project duration.
- The practice of prefabricating parts of structures rather than building them piece by piece may often accelerate a programme.
- In an information technology project the use of standard equipment and existing software offers a means of reducing project durations.
- In any industry the sub-contracting of work to specialists may be more expensive, but may speed up work especially if the sub-contractor has highly skilled and well-equipped teams.

The time/cost relationship for the project as a whole

Early in this chapter Fig. 40 showed a curve which represented the way in which the cost of an activity depended upon its duration; later this was assumed to be a straight line in order to simplify calculation. It is of interest to draw a corresponding diagram for the project as a whole, as shown in Fig. 44. Data for this diagram are obtained from Table 8. This is followed by Fig. 45 which now also includes the daily overhead charge of £22, and shows that the project has an overall minimum cost of £856 at a total duration of 18 days; this is the same result as was achieved in the tabulation, but also indicates that the total cost does not vary very much if the project duration is within a day or two of this either way. It is obvious that the numbers used in this case have been chosen to keep the arithmetic simple. In practice it may be found that this curve does not have a minimum value intermediate between the normal and crash durations, and that the lowest cost is at one or other of the extremes; it would still be useful to know whether it is at the normal or crash duration.

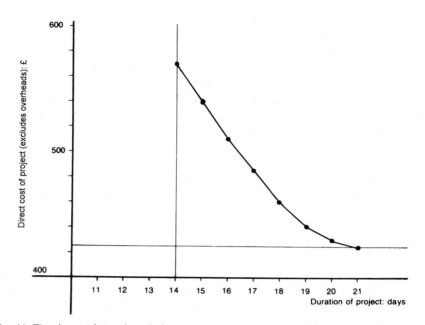

Fig. 44. Time/cost relation for whole project for the same example (based on direct costs only)

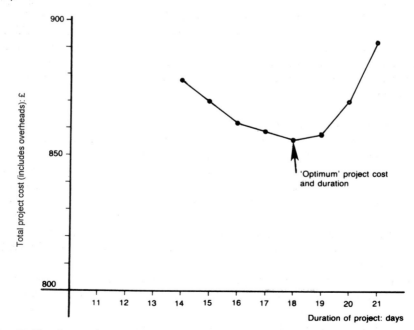

Fig. 45. Time/cost relation for whole project with overheads included and showing the optimum project cost and duration

THE INTERACTION OF QUALITY WITH COST AND TIME

Much of this chapter has been devoted to the calculations that can be made relating to the way that time and cost interact. There is no direct equivalent calculation for the relation between quality and either time or cost, but it does need to be examined. It has now been said a few times in this book that once quality standards have been specified in a project they should not thereafter be negotiable as a means of reducing either time or cost, and because of this there is little scope to take advantage of the relationship in the processes of managing project implementation. There are still two aspects which must be considered

(a) the impact of these interactions on the briefing and design of the project
(b) the management of the project if it shows signs of failing to meet any one of the criteria of quality, time or cost.

Quality and cost at the briefing and design stage

One of the main decisions to be taken at the outset of a project is the general standard of work and performance that has to be set. This then has to be interpreted in terms of the specifications of materials, equipment, workmanship and performance. One of the fundamental questions which has to be tackled at the outset of a project is the determination of the expected life for the facility being created. This topic is more fully discussed in Chapter 12 in relation to project investment appraisal where the concept of lifetime costing is looked at in some detail. This concept states that in selecting, for example, a piece of equipment not only the initial cost should be considered, but also its expected and/or required useful life, its operating cost, its repair and maintenance costs, and of course the quality of product or service it will give. Putting this into terms of specifying components or materials for a project can be illustrated by a few practical examples.

- When designing fire-fighting equipment for an off-shore oil platform it is important that reliability is given high priority, but the life requirement would be relatively short. Operating cost is not at all important, but because of the difficult environment the required maintenance operations should be minimised. Initial cost would be a significant but not overriding factor. It is likely that the specification for this equipment would call for a high standard, would be fairly expensive, and adequate time would be given for it to be designed, manufactured, installed and tested all to a high standard of performance. Given that the cost of failure could be extremely high this could not be regarded as a low-cost item.
- Non-recoverable items incorporated into space projects will similarly have a requirement for very high reliability, but will mostly have a very short life. In such cases it is more important to have several stages of back-up rather than that there should be no deterioration after prolonged use.
- Pumps used in the continuous operation of chemical plants are often set up

with several units in parallel so that any one failing will not endanger the process. In this type of application wear will be considerable, and hence running life is important, but need not be as long as the expected project life, as the pump can easily be replaced at a reasonable cost.

- The specification of materials in buildings presents an interesting case, for example timber windows and doors. Where top-quality materials are required for aesthetic reasons the decision may be simple, but in more utilitarian situations it needs thought. Hardwood is much more expensive than softwood, but the latter will need much more maintenance over its lifetime. On the basis of life costing a case can often be made for the use of hardwood, provided of course that it comes from a sustainable source. It is seldom used however because there is greater emphasis on the initial project cost rather than on its subsequent maintenance. This is partly due to the fact that the initial outlay comes from one pocket of money and the running and maintenance costs from another, possibly even from a different organisation. This is certainly true in speculative house-building where there may be pressure to keep building costs low and little regard is paid to long-term maintenance.

Good project management would look objectively during the briefing and design stages at the ways in which quality interacts with cost in particular, but also in relation to time. It now remains to discuss how quality influences time and cost when it comes to the implementation stage.

Spending time and money to ensure quality

At the stage of implementing a project it is possible that a number of quality issues will arise. There are the occasions on which it will be found that an individual item of work is not up to standard; this is almost inevitable unless very rigorous precautions to avoid defective work and goods are taken. The objective of having 'zero defects' is a very sound one but is rarely achieved in project work at present; it remains an important target for the whole discipline of project management. When defects are detected a number of options are available, depending upon the circumstances. If it is simply a defective component then it is likely that it can be replaced by a new one, probably at the supplier's expense. A defective component that has been incorporated into some other work presents an added problem, namely who has to pay for the dismantling and rebuilding that has to be done? There will also be the delay caused by the reworking. When defective construction or installation has been found it is often difficult to find a totally acceptable solution. Most building professionals will have come across such problems: concrete with a poor surface finish is extremely difficult to repair properly; concrete which is defective can be broken out, but it is almost impossible to make a sound connection between old and new concrete. A hole which has been dug in the wrong place has to be refilled but with what material? Refilling with excavated material cannot be compacted well, refilling with concrete may be too rigid and lead to other problems. In many of these fault situations in construction the time and cost of removal may well exceed the time and cost of the original construction, with the further addition of reconstruction which may easily be more than the initial cost.

The points related above are the cause of a great deal of pressure being brought to bear in the construction industry to accept some items of work which are below the specified standards; the argument is that the delay caused by removal and reconstruction is too great to justify the action, and in many cases the work is said to be 'good enough anyway' for the purpose. The fact that it also saves someone considerable expense is not often used as an argument but it may frequently be a hidden reason. One of the ways of counteracting the argument of 'do nothing' is to insist on building into the programme not only a contingency sum to allow for unforeseen work but also a contingency time to allow for the correction of defective work. This may present a possible way forward in the struggle to achieve zero defects, since it removes the risk of causing project overrun by insisting on all work being up to specification.

The approach of manufacturing industry to prevent all defective work could ultimately be made the norm in project work if enough effort is put into planning, supervising and the control of materials and components. If this were achieved it would indeed be the case that there is no trade-off between cost and quality, because it will be found that good quality always pays off in the long run. It may be difficult to persuade all parties to a project contract that this is the case, but from a total project point of view it can be seen to make sense.

Learning Resources
Centre

9 Project implementation

At this stage of the book many of the technical details have been discussed — this chapter will explore the actuality of applying project management principles to the management of a real project. This aspect of project management has been addressed by a number of writers, including Fangel who has written much on 'project start-up' (e.g. Fangel, 1985), and there are some joint papers on 'project implementation profile' by Slevin and Pinto. It is somewhat difficult to generalise on the subject, and the examination of any one case will of necessity omit many aspects of importance. The approach here therefore is to relate one case (which was introduced in Chapter 4) and which includes a number of different operating industries, and then amplify this by a series of additional comments which might apply in other situations. The chapter goes on to discuss the management of resources, a topic which has a special importance in project management.

A MULTI-DISCIPLINARY CASE STUDY

Photo Products Ltd (referred to as PP Ltd from here on) has been operating in the photographic industry for some years, but has confined its activities to the manufacture of film, processing equipment, printing paper and associated consumables. The company has decided to enter the market for popular equipment with an innovative camera and has satisfied itself that the proposal is viable; a decision to proceed has been taken, and the project should now be put into action by the establishment of a project team. (Note that study of this case is being taken up part way into the project, i.e. after the project appraisal has been completed; this is in itself an important task which is the subject of Chapter 12, but is not pursued here.)

Project strategy and organisation

In any project a general strategy should be established at a very early stage, determining broadly what has to be done and by whom. This should include all the major factors such as design, finance, control, approvals, risks, human resources, environment, procurement and information systems. One of the first tasks for PP Ltd in the case of their project is to decide the form of project management organisation it intends to operate, and the way in which the project is to be structured. This is sometimes referred to as 'organisation breakdown structure' (Turner, 1993; Thompson, 1981). It covers a number of decisions about who will be responsible for each main part of the project e.g. market research, product design, facilities procurement and all the other tasks indicated in the network shown in Fig. 46. In this case a decision is made to release from the plant for the duration of the project a PP Ltd staff engineer who is given the title of project manager, in control of a project team. It is important that the person appointed has very clear terms of reference, and in this case the terms include reporting directly through the chief executive to the board. The post carries delegated authority to control all aspects of the project, subject only to seeking approval from the board for contracts in excess of £1 million. Other organisational decisions would concern the appointment of product design consultants to work with a newly recruited in-house team, the commissioning of a market survey, appointment of architects and engineers for the new building, and briefing the existing information technology group to work on systems for the new plant and product. Many other decisions would be required, but these are sufficient to indicate the organisational structures that would have to be established in such a project. It is vital in any project that the methods of communication are properly designed and established. A simple but thorough approach to this is given by Woodward in *Science in Industry, Science of Industry*, the essence of which is that all communication should be valid, relevant, concise and clear. In project work the medium of communication may be by drawing, written word, word of mouth or by electronic means, and should be recorded where appropriate. This is discussed further in Chapter 14.

Other project management structures

It should be noted here that many other alternative structures could have been put in place to handle the project management; ranging from giving responsibility to the existing plant manager (but this would mean a part-time project manager with possibly conflicting demands on time). Another means would be to give responsibility to a project committee of existing staff (but this runs the additional risk of divided authority and loyalty). An alternative possibility is to appoint an external consultant who can be full time, and can supplement resources when needed, but will not know the company and may take some time to get started as a result. Whichever means is used to appoint the project manager it is essential that a project team is set up to carry out the detailed work; the team may comprise both full-time and part-time members who report to the project manager. The whole subject of people in project management is covered more fully in Chapter 10, including examination of the skills and experience needed, and the way in which teams operate.

Project outline document

Given the existence of the project team, one of its initial tasks is to establish its method of working, and obtain formal approval of this. One method of doing this is

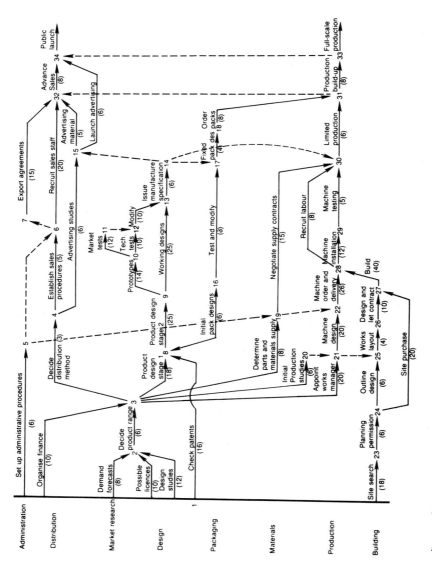

Fig. 46. Arrow diagram for project Photo Products Ltd — new production facility for new product (durations shown in weeks)

to prepare a project outline document which sets out the main features of the project including an introduction to the industrial context, a business plan, the objectives and scope of work, an outline technical description, the project organisation, the project time-scale, and a risk analysis. This document, once approved, would be expanded to form a reference manual, sometimes referred to colloquially as the 'blue book' or the 'project Bible'. This would go into much more detail, and would be kept up to date throughout the project, noting any changes and their dates so that it can also form the basis of a project record as the work proceeds. The list of items to go into this expanded project document serves as a very useful basis of discussing how the project might proceed. Turner (1993) gives a very good description of what he refers to as a project manual to cover this aspect of the work.

Project scope and organisation

As has already been discussed in Chapter 4, it is essential at the start of the project to set down what it includes and where responsibility for each part lies. For PP Ltd the scope might be stated to include the following areas.

- Market survey, not only to determine the potential size of the market, but to obtain detailed information to assist in the design of the new products.
- Carry out initial product designs and assess their means of production.
- Determine the most appropriate means of funding the project.
- Investigate the present position on patents.
- Design and estimate the cost of the means of production.
- Investigate the means of distribution and marketing.
- Identify suitable sites for the manufacturing plant, taking account of site availability and cost, location relative to the market and the means of distribution, labour skills available, financial assistance, technical support and several others.
- Design of the factory building.
- Planning and other consents.
- Forms of procurement for buildings and plant (contracts).
- The construction of buildings.
- Plant installation and commissioning.
- Determination of materials and components sources.
- Information and control systems for the plant.
- Staff recruitment and training.
- Technical documentation, back-up service and publicity.
- Product launch.

Project programme

The techniques of management of time in a project were explained in some detail in Chapter 5, but it remains to discuss how the techniques are applied in practice. In a project such as the one under discussion it is likely that an outline network of the whole project would be drawn at the outset to help management to make major decisions; it is probable that a broad programme would be needed as part of the go/ no go decision, the point at which we took up this case. Fig. 46 shows a form of this outline network, and it can be seen that this embraces the scope items that were listed above. It is possible that some readers might wish to modify this network, and this is exactly the way in which an outline programme should be used, namely as a means of evolving a more detailed and more representative working programme. At this point there are two possible alternatives, the first being to expand the outline network by breaking each activity down into a number of component parts and adding others. This would form a comprehensive complex network with a large number of activities; in itself this presents no major problem but for some projects the second alternative may be preferable. This second alternative is useful if the project lends itself to being divided into a number of separate areas, either physically or organisationally distinct, then each of these areas can be represented by a sub-network. Such sub-networks, e.g. Fig. 47, can stand on their own for detailed work, and can also be coordinated with the whole project through a master network which is essentially the outline network updated as necessary. This allows each sub-network area to be planned and controlled by the people directly concerned, and the sub-networks can then be put together, usually with the help of a computer package, in order to manage the interactions between separate areas of work. It should be noted that in Fig. 47 there are several design activities which run in parallel, and which could all be represented by arrows all starting and finishing at the same pair of events. While this is logically correct it may cause a problem with some software packages which require each activity to be identified by a unique pair of event numbers. For this reason a dummy activity has been inserted in several of the paths to achieve uniqueness; but in each case it has preceded the activity rather than followed it. This is because where free float arises, it is located on the last activity before a critical event, and there is little point in having free float on a dummy.

In the case of the PP Ltd project it would make sense to have separate sub-networks for many of the work areas, e.g. the product design, the search for and acquisition of a site, the design and installation of the production systems, the design and construction of the factory building, the determination of marketing and distribution systems, the information control system design and installation, and several others. Fig. 47 shows an example of one typical sub-network, and Fig. 48 an indication of the way in which sub-networks can be put together into a master network. In this figure each of the separate sub-networks has been simplified to a small number of arrows, which represent the main areas of work in that sub-network. These arrows intersect at key events, which can be cross-linked to corresponding key events in other sub-networks. These are often referred to as 'interface events', for example Fig. 48 shows that the production systems design cannot commence until the preliminary product design is complete, and in turn has also to be carried out before design work can commence on the buildings. Each of the sub-networks can be managed by a separate group of people, but all sub-networks can be linked in order to give the project manager the information needed for overall project control. It may now be useful to look at the way of approaching the initial drawing of one of the sub-networks.

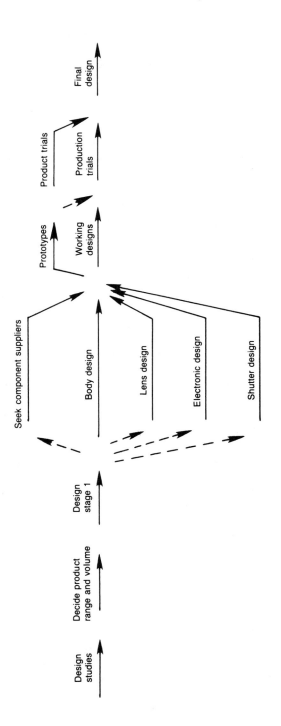

Fig. 47. Sub-network of product design for Photo Products Ltd

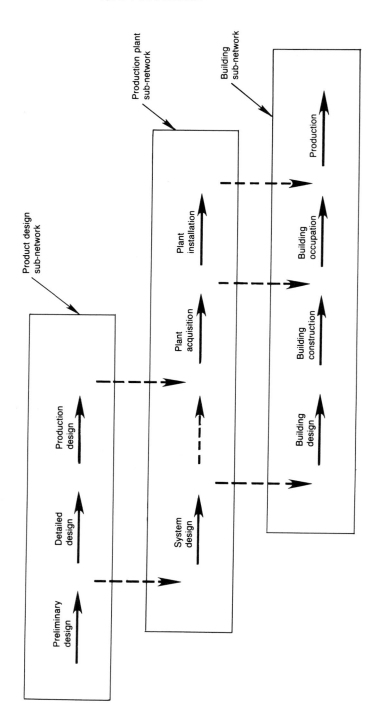

Fig. 48. Linking together sub-networks into a master network for the Photo Products Ltd project

Drawing a sub-network

In Chapter 5 it was stated that an early decision to be made is the level of detail to which the project should be broken down. If day-to-day control is required then it will be necessary to have most activities with a duration of only a few days. For example, in the network for the building construction for PP Ltd it would not be satisfactory to have a single activity 'build foundations' since this would last many days, cover a wide physical area and involve several different resources. Perhaps more importantly it should be possible to start work on the construction of columns and walls as soon as a reasonable length of foundation has been completed. It is not easy with an arrow diagram to show the overlap of two jobs such as foundations and walls, and even with a precedence network the use of lead and lag times is not infallible. This potential problem can be explained as follows: Fig. 49 shows two activities, foundation and walls, with a lead time of 4 days and a lag time of 6 days. These are intended to ensure that four days after foundation work is commenced it will be possible to start building the wall, and three days after the foundation is completed there will remain three days of wall building to complete. If all goes well then this representation is broadly correct, but the fact that four days have passed since the beginning of foundation work does not guarantee that enough of it, and the right bit of it, will be ready to accept the building of the first part of the wall on the fifth day. The only way to ensure that the correct progression of work is represented by this network is to break the work down into separate physical areas of work and indicate the full logical sequence of activity, as shown in Fig. 50 which is sometimes referred to as a 'ladder'. This assumes a sequential approach to the foundation construction, starting with foundation A and then moving on to foundation B as soon as foundation A is complete. This way it is possible to commence wall A and continue with foundation B at the same time. Similarly when foundation B and wall A are complete then wall B can be built. The whole length of the structure can be divided into any number of sections in this way, but it would be usual to make each section big enough to accommodate a gang working on it, say two or three bricklayers working on the same section of wall. For the PP Ltd project Fig. 51 shows how a precedence diagram would represent the overlap of the foundation and walls in the case outlined above.

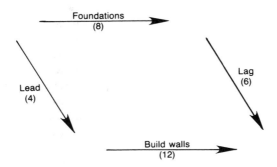

Fig. 49. Ladder arrow network for overlapping activities

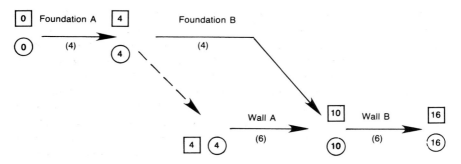

Fig. 50. Ladder arrow network showing overlapping activities

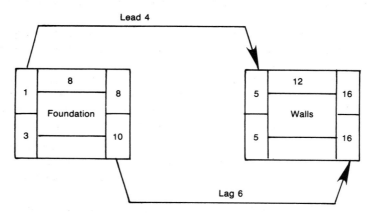

Fig. 51. Overlap of foundation and wall activities using precedence network

How to set out the network

At this stage reference should be made back to Chapter 5 which sets out the steps to be taken in drawing a network. These are reviewed here with reference to the PP Ltd project. The drawing of a network forces the planner to go mentally through the project, thinking about it step by step. The best approach is to set this down in the form of a network in logical sequence, either as an arrow or precedence diagram on paper, or it may be in the form of direct input to a computer (see Chapter 14). There is no need to write down a list of the activities, and indeed the writing of a list should be avoided because it implies a sequence which may not be totally correct. The network should be drawn initially without entering the duration times, as the principal objective is to create a network which is a model representing the essential logical sequence of the activities. The use of the word 'logical' is quite specific, and means showing two activities in sequence only if it is necessary for the first to be complete before the second can commence. Care should be taken to ensure that the sequence set down does not simply follow normal practice or a preferred way of operating; it is important to question each dependency that is drawn to make sure that it is absolutely necessary.

A convenient layout in the case of arrow diagrams is illustrated in the PP Ltd master network in Fig. 46. This is by no means an essential concept, but it is sometimes convenient to put all the activities relating to each technical aspect or section of the work in one row or 'band' of the diagram, as indicated by the titles in the left-hand column. This makes the reading of the network somewhat easier especially for those who are only concerned with a limited part of the project.

The concept of unlimited resources   Another pitfall is the temptation to set a series of activities in sequence because there is likely to be only one unit of resource available and it is known that only one activity at a time can be carried out. While this may be true it is often the case that no specific sequence is appropriate, and indeed there may be ways of duplicating the resource if it is important. The way to overcome this is to make a temporary assumption that there are unlimited resources available and that unless there are individual sequential tasks then no regard should be taken at the outset of competing demands for the resources. At a later stage when potential timings are known and a working plan is being set down, decisions on resources and sequences can be made.

Analysis first, plan of action last   If designers are tackling a project it would be usual to start with the brief, collect information, analyse the problem, consider alternatives and finally come up with the proposed design. In a rather similar way the project manager should look at the project as a series of sequential steps.

(*a*) Set down the essential logical dependencies, without regard to time, resources or preferred methods. This corresponds to the analysis stage of a design.

(*b*) Work out activity durations, and then calculate the minimum project completion time.

(*c*) Determine resource needs, and check the impact of resource limitations on the programme.

(*d*) If there is an overrun, examine ways to overcome the problem, possibly by spending money (see Chapter 8).

(*e*) If there is time to spare, see where this can be used to smooth out the demand for major resource items.

(*f*) Search for potential problems and seek to prevent or solve them.

(*g*) Prepare a 'plan of action' which is a statement of what is intended. This corresponds to the working drawings prepared by the design team, and is effectively the design for the project implementation. This plan of action may be in the form of a network, a bar chart, a series of dates in a schedule, or perhaps as a computer printout of any of these. The preparation of this working plan is often referred to as scheduling.

(*h*) Once the project implementation has started the actual performance in terms of time, cost, and quality has to be compared with the schedule, and if any deviations arise action must be taken to overcome them. This is in accordance with the general management cycle of 'plan–measure–control–replan' that was described in Chapter 2.

The above section is really pointing out that in a design the engineer or architect does not start with the working drawings and hope that these can incorporate all the thinking processes of the design. The drawings evolve after a long process of analysis, evaluation, and comparison of alternatives. In the same way the schedule or work plan for the project manager is the output and result of the planning process. A common mistake made on some projects is to take an existing bar chart, perhaps prepared as a statement of 'wish' by the client and use it as the basis of preparing a network; all that happens by this step is that the mistakes and preconceptions of the client become built into the plans of the project manager. All the project manager needs in order to commence the planning process is the factual information of the technological content of the project, any major restraints and key target dates such as partial hand-over, and a large clean sheet of paper!

The drawing of networks These may well require large pieces of paper, and planners should not be put off by the critical remarks about 'wallpaper'. Networks can be directly input to a computer, but it is difficult to obtain an overall view of the network as it proceeds and this is better done on a sheet of drawing paper. The use of a computer to carry out all the calculations is now very common and helpful, and the printed output is generally very suitable as is described in Chapter 14. Once the data has been fed into the computer there is relatively little need to refer to the large network drawing, except for the purpose of amending the logic, which should not happen very often. The network drawing will rarely have planned dates on it, as these are likely to change and are better on the computer print-out. There is no doubt that the process of drawing a network is in itself a valuable step; it has already been stated that the network is really a model of the implementation/construction process, and that one of the main uses of a model is as a means of learning about the project. The actual drawing of the logic diagram forces the planners to think their way through the project in a logical sequential manner, and at the same time learn about its detail. It is sometimes said that even if the network is thrown away (or as more often happens is put away in a drawer), then it will still have been worth preparing because it has forced the planners to think about the project in detail. One of the problems of one-off work which characterises projects is the total lack of prior experience of the particular project, and this can be partly substituted by the exercise of preparing the network.

At one time it was thought that the whole of project management revolved around the use of networks, and many of the early books and courses concentrated on this. It is now generally acknowledged that there is far more to the subject than the simple preparation of a diagram. This does not mean however that the network can be eliminated, and in fact even if it is not drawn it does exist in a hidden way in the form of a series of dependencies which are intrinsic in the work. Another fashion which has come into being is the use of computers, and while these can be very useful, especially on large projects, they are not always essential. It is perhaps the enthusiasm of the computer experts which has created the impression that computer use is essential in project work, and it is noticeable that most of the stands at project management exhibitions are concerned with hardware and software. In many cases it is possible to feed the logic sequence directly into the computer, but this does have its limitations and it will usually be preferable to start planning with a sheet of paper and a pencil. As was stated in the earlier listing, the initial network should be drawn in terms of logic alone, without inserting expected durations, and without having regard for resource requirements.

Time calculations

Once the outline logic has been agreed the next step is to calculate the project timings. The first decision will be to select a time unit appropriate to the project, and a very rough rule for this would be to select a unit such that the project as a whole has a duration of the order of 100 units. This would mean that a project of 2 years' duration would be planned in weeks rather than days, while one of 10 years total might be planned in months, and one of 3 months duration would be in days. This is an extremely rough guide, and certainly the time unit chosen should be one which is in common use. One exception to this is the use of a 'work-shift', especially when a project is working on 2 shifts a day, 5 days a week, which then gives 10 time units a week and a simple decimal translation from basic time units to weeks. It is generally inadvisable to use more than one time unit in any one project; although there are some computer packages that can handle any combination of hours, days and weeks etc. it is possible that human project managers would find it confusing. In the case of the outline or master network for the PP Ltd case the time unit used is a week, and the same unit is also used in the more detailed sub-networks. The next and rather more difficult task is to estimate the durations of each activity in the network, and this will usually involve examination of past performance on similar tasks elsewhere. When there is no prior experience available it may be necessary to break the task down into elements and build up a composite time, as might be done by a work-study expert. The approach should always be to assume that the task is tackled 'in the normal way', i.e. taking no special measures, so that the duration figure used represents the economical and most practical way of doing the work. Chapter 5 gave in detail the method of carrying out the calculation of earliest and latest event times and hence the minimum duration of the project. This could also be done by any of the project management software packages.

Consideration of the project duration   It will usually be the case that the calculated project duration is either longer or shorter than the time available. If the time required is longer than that available then clearly some extensive rethinking is needed, since at first sight the project appears to be impossible. For the present argument it is assumed that the logic is correct and that it has been checked; there is thus no possibility that the network can simply be compressed by revising the logic. If the diagram truly represents the project then some real change must be made to the project to achieve acceleration. This change could take one of many forms, possibly amending the scope to cut out some part of the work or substituting something simpler, such as PP Ltd taking over an existing building instead of designing and constructing a new one. A major change might be made to the technical methods being used in the project, e.g using existing software rather than writing new for the materials and components purchases for PP Ltd, or using prefabricated construction for some of the building. It must be emphasised that some real change must be made and that the temptation to 'reduce all durations by 20%' must be resisted. If time cannot be reduced by scope or technical means then it may be possible to accelerate some activities by extraordinary means such as significant overtime or double shift working. This means of reducing time, usually at some cost, has already been discussed in detail in Chapter 8. It may be necessary to carry out several iterations of the network time calculation, and this is where computers are very useful; once the network detail has been entered it is a simple matter to change some of the activities and/or their durations and obtain a revised project completion time almost instantaneously.

When the calculated project time is shorter than the available time the problem may at first appear to be simpler. If it is shorter by a significant margin then there may be relatively little difficulty, and indeed some consideration should be given to planning to complete the project early. In other cases where there is little excess time available, what has to be remembered is that the original network was drawn on the assumption of unlimited resources. In most practical projects this will not be the case and some levelling out of resource demands will be necessary, with the consequence that some activities which had originally been shown in the network as parallel will now be sequential and may result in the project duration being extended. If this extension still falls within the allowable time then all is well, but if it causes the project to overrun then further thought will be needed. Again there must be an iterative approach to matching the time required to the time available and this should be done at the planning stage rather than waiting until some part of the project is falling behind schedule. Once all the adjustments have been made to the plan it can be agreed by all parties to the project to be a basis on which to proceed, and it is issued as a working plan or plan of action. At this stage the project can commence.

RESOURCE LEVELLING AND RESOURCE ALLOCATION

Reference was made above to the fact that in practice there will be a limit to the availability of resources on most projects, and some form of resource allocation and resource levelling will have to be carried out as a deliberate action. It is not satisfactory for a resource to be used by the group which makes the loudest call for it, since this may mean that a critical task is delayed unnecessarily. The subject of resource allocation is worthy of more detailed study, as is set out in the following paragraphs.

Types of resource

It is possible to classify resources for most projects under four main headings, and then to put these into two broad groups. The first group is that of manpower and machines. These are referred to as 'time-based' resources, in that it is the time spent working of these two resources that is important. They cannot be saved up from one day to the next, and it is not possible to double the amount of work done by one unit of resource on any one day. The second group consists of money and materials. These are called 'consumable' or 'pool' resources, i.e. those which are used up. Each unit can only be used once, but if not used on any one day can be saved up for use on another day (provided that it is not perishable).

With consumable resources it is possible to organise their availability to suit whatever project plan is wanted, even if this means an erratic use, by arranging delivery in a non-uniform way, or stock-piling them in advance of need. In contrast the use of time-based resources has to be planned carefully to make sure that each unit is put to good use every day; if any unit is not used on a particular day then that day's work is lost but still has to be paid for. Furthermore, if there are not sufficient units to meet the demand on any one day then one or more tasks will have to be delayed. It can therefore be seen that it is not necessary to think separately about all four different resource types in the planning of work, simply those that are 'time-based'. It should always be possible to make consumable resources available for any programme that is required. This is not to say that such resources never create problems, but this is not due to their inherent nature but to the fact that they have arrived late, or not up to standard, namely a failure of supply. The main problems then of managing resources is to decide how many of each type of time-based resource units should be made available to the project, how and where these should be allocated to tasks, whether and when the number of units should be increased or decreased, and the extent to which a restriction on the number of units made available will shorten or lengthen the project duration. The best way to explain the approach is to work through a numerical example.

Fig. 52 shows a histogram of the needs for a particular manpower resource on a simple 10 week project, based on the assumption that each activity is planned to start at its earliest possible timing. The question to be answered is whether it is possible to complete the project in 10 weeks with only 2 units of the time-based resource available at any time. A quick inspection of the diagram will show that the total demand for the resource is 20 worker-weeks, giving an average demand over the project duration of 2 workers. This implies that the project will not be delayed by this resource limitation, provided that there is sufficient float to allow some of the activities to be delayed to spread the demand. This conclusion is not correct however, because closer inspection shows that there is a 'spare' resource unit in week one and as this time-based resource cannot be saved up it will be lost; consequently

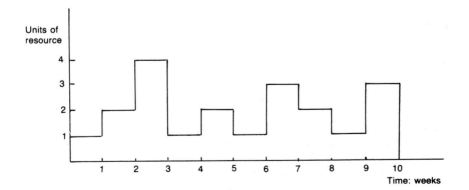

Fig. 52. Resource histogram

the project cannot catch up this loss and will overrun. A similar conclusion could also be reached from the fact that the project requires 3 units of the resource in week 10, but as there are only 2 available there will inevitably be a delay on the last activity.

Example of the impact of resource limits

Fig. 53 shows the network for a simple project, indicating the durations of each activity together with the calculated earliest and latest event times. A requirement of this project is that 1 unit of resource X is needed on activities 1–2, 1–8, and 3–4, and 1 unit of resource Y is needed for activities 7–10 and 8–9. While this type of problem can in most cases be handled quickly by computer, it is important to understand what steps are taken to solve the problem, which may be stated as: is it possible to complete this project on time with 1 unit of X and one unit of Y? The way to tackle the problem is to draw a bar chart to scale of the activities involved (shown in Fig. 54), and take each activity in turn as is done in the paragraph below. The process is much more easily explained in the lecture room using a chart, and inevitably makes rather tedious reading, but it is worthwhile following the steps through carefully by referring to the diagrams. Once understood it is logical and simple, and rarely has to be done manually in any case.

First we plot on to the bar chart the earliest and latest timings of the three activities which need resource X, and then ask the question of where should X be initially allocated. Given that activity 1–2 can start as early as any other X activity and that it is also critical, it can be seen by inspection that X must be used initially on that activity, and will be in use there until the end of week 5. The next decision offers two alternative allocations, to 1–8 or 3–4, and both of these must be examined. If X is used on 1–8 it will be used there for 3 weeks and will not be available on 3–4 until the

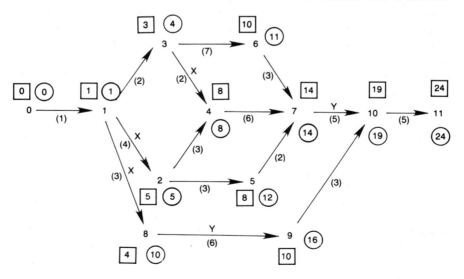

Fig. 53. Arrow diagram for resource smoothing exercise

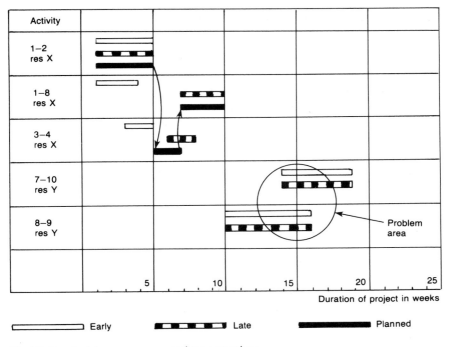

Fig. 54. Bar chart for resource smoothing procedure

beginning of week 9, which is far too late and will delay the project. The alternative of allocating X to 3–4 at the start of week 6 would allow 3–4 to complete by the end of week 7, which is 1 week before the critical date; moving resource X across to 1–8 at the end of week 7 would just give enough time for the execution of 1–8 over weeks 8, 9 and 10, which is the critical latest event time for event 10. It has therefore been possible to carry out the three activities needing resource X within their range of earliest and latest timings and not delay the project. The next stage of the exercise is to carry out the same process for Y and consider activities 7–10 and 8–9.

Referring again to the bar chart of Fig. 54, activity 7–10 can be seen to be critical and must therefore be allocated resource Y at the start of week 15 if the project is to be completed on time. Turning now to activity 8–9 it would appear from the network that this could be started any time after week 4 and hence be completed well before week 15 when resource Y is needed for activity 7–10. However, it must be remembered that activity 1–8 has already been delayed to its maximum extent in the manipulation of resource X, and will not be completed until week 10. This means that there are now two activities which are both critical, which overlap by 2 weeks, and which both require resource Y; clearly there is a problem which needs some thought. Faced with this situation the best approach is to write down the alternatives which are possible and to compare them before making a decision; typical suggestions from a project manager are listed below.

(a) Work overtime on the activities requiring X or Y. This is valid, but the idea could be extended to the concept of reducing durations anywhere on the critical path by any means. For example the problem could perhaps be solved by a big increase of labour for the final activity 10–11, reducing it by two weeks, and allowing 7–10 to use resource Y in weeks 17 to 21 inclusive.

(b) Allow the project to overrun by 2 weeks. This may seem to be undesirable, but is the outcome that often arises.

(c) Start 2 weeks early. This will often be acceptable, but is not always possible, especially if the planning is carried out too late.

(d) Double up on resource X. The objective was to avoid this, but it might be the best of a number of undesirable alternatives.

(e) Double up on resource Y. Again this is undesirable, but may be the best available.

(f) Change the project method. On the assumption that the original network was correctly drawn it should not be possible to change the logic without making some form of physical change in the project itself, perhaps in the design or procurement procedure.

There may be other alternatives and these should all be listed and examined carefully to find which of them is least undesirable. In this way the project can be executed in the most acceptable way. The time at which this exercise is carried out is important, as exemplified in the following possible scenario. Consider for a moment that in the above project the network has been correctly drawn but without the resource-levelling procedure; work on the project has been carried as far as its half-way stage, i.e. 12 weeks, with all activities completed in their planned times, apparently a generally satisfactory situation. In the absence of resource-levelling for the whole project it has been tackled on a month-by-month basis and resource X has been allocated in the way shown in Fig. 54, and has completed its work at the end of week 10. Looking 4 weeks ahead it can now be seen that there will be a problem with the overlap of activities requiring resource Y, namely 7–10 and 8–9. Reference to the previous list of alternative solutions now shows that some of them are no longer available to us. Alternatives (c) and (d) are not possible, we cannot now start early and cannot double up on X. Option (a) — reduce durations on the critical path — is still open but its scope is much reduced because so much of the project is complete or largely committed. Option (f), i.e. changing design is almost certain to be no longer possible, leaving as realistic options only (b) and (e). Doubling up on resource Y is still open to the project manager, but may be difficult or expensive. Unfortunately what happens in many cases is that the approach seems to be 'let's wait and see and hope for the best', with the result that the project runs late, namely the last remaining one of the six options in the list above.

Lessons to be learned

What has really happened in the preceding example is that the float in the project has been used up too early. (For a detailed discussion of float refer back to Chapter 5.) In the original network in Fig. 53 it can be seen that there are 6 weeks of total float along the path through events 1, 8, 9, 10, and its use should perhaps have been more carefully considered. In the allocation of resource X all the float in this path has been used on its second activity. The lesson is that float is a very valuable commodity, and should only be used with care; what often seems to happen is that it is simply allocated by default, and the early participants in a project just carry out tasks using a common resource in any sequence they wish. Clearly the best use of resources should be properly planned. It is worth noting that sometimes one of the worst culprits in delaying the early stages of a project in this way is the client.

A second important lesson is that the 'satisfactory' performance of the project at the half-way stage in the preceding example is more apparent than real, and this is a situation which arises often; everything is seen to be going well in a project perhaps as far as the 75% stage, and then suddenly it starts to go wrong. What has happened is that the periods of float on various paths have largely been used up, leaving no scope for levelling out resources. In reality there is a real limit to the availability of the resources, and some activities just have to wait, even though they are now critical; the consequence is inevitably that the project runs late, with a disgruntled client, a harassed project team, and an expensive and tedious claim for delay to be settled. How much better it would be if all this effort was devoted to avoiding the problem in the first place! Good planning before the project 'gets off the ground' can repay handsomely.

The difference in approach between British and Japanese industry is of note in this respect. In cooperative ventures it has often been found that British managers become impatient with the time taken by their Japanese colleagues to talk around a project before 'getting down to the real work'. Once the talking is over the Japanese seem to be able to get the project completed more quickly, and without delays. It has long been a tradition in the UK contracting world to demonstrate enthusiasm by setting heavy earth-moving machines to work on the first day after the award of a large contract; This may be a grand gesture to impress the client, but there has been more than one project in which a large hole was excavated in the wrong field!

Resource domination

One additional aspect of resource management is the extent to which an individual project is dominated by the question of resource demands. At one extreme a project which is to be undertaken totally by one person is likely to be greatly dominated by that fact, as activities will have to be undertaken as a long series. This does not mean that there is no need to plan, as some of the activities will also have external constraints. Such projects can be thought of as being 'resource dominated'. At the other extreme there may be urgent projects where every possible resource need is met in order to ensure rapid completion, for example the transfer of manufacture from one production line to another involving loss of production during the transfer. Such a project may be referred to as 'time dominated'. In between these extremes will be a number of intermediate cases, even within one industry. For example, the construction of a multi-storey building in an industrial area will be able to call upon the services of a large number of specialist suppliers and sub-contractors who can be mobilised at short notice, machines can fairly easily be hired in for short periods, and in many ways a variable resource demand can be accommodated. By comparison the same building in a remote location such as an off-shore island, cannot so easily be served and must be more self-sufficient; this also means that such a site must try to make more steady (or 'level') use of its plant and people since their numbers cannot be easily varied. It can therefore be thought of as being more resource dominated than the urban project, and will probably also take longer to complete. One particular example of an absolute resource limit is the case of the fitting out of an off-shore oil platform, where the number of people working at any one time is limited to the number of beds in the accommodation module. The mix of different skills can be varied with careful planning as the changeover is made each one or two weeks, but the total number can only be increased by the use of a separate accommodation unit, which would mean a very sudden and large increase in the manpower, i.e. such projects are heavily resource dominated.

Work breakdown structure

Work breakdown structure (WBS) is not the same as the organisation breakdown structure which is essentially concerned with the responsibilities of the various groups or organisations, and the interactions between them together with the necessary communication and decision-making. The breakdown of work is usually more detailed, and may be divided by a number of different factors.

- The content of the work, e.g. a particular technology such as the design and provision of equipment to carry out the manufacturing processes for PP Ltd. Similarly the design and installation of a management information system for the plant would be an identifiable package, as would the recruitment and training of the workforce.
- Physical location might be the basis of defining a work package, for example work on one site or in one building or one department in an office. The building construction might itself be divided into a number of work packages, such as ground works, structure, finishes, services, and external works. Alternatively where there are several separate buildings the packages might each consist of one total building; there is no absolute rule about the division into packages, it will be a matter of what suits each project.
- Another basis of separation is by separate trades, especially where there are different traditional working practices, some groups having highly structured working agreements while others have a more flexible pattern of self-employed workers.

Work breakdown can be established in whatever way suits the project, but it has a number of common threads. The size of a work package is kept to that which can easily be controlled by one person or a small group, and often within a specialised subject. This makes it easier to set and achieve targets, to coordinate resources and to provide project leadership. In Chapter 2 it was said that one of the problems of project work is that its unique nature means that a high degree of control must be exerted at each stage and level of the project; this is made easier by the concept of work breakdown. In some cases on large projects one package may be very large, and may in turn be broken into a number of smaller packages, in a form of nesting. This whole topic is again one which is covered in depth in *The Handbook of Project-Based Management* by Turner, and is also discussed by Badiru (1988) in relation to manufacturing and high technology operations.

Breaking a project down does not mean that each package is independent of all the others, and it is necessary to ensure that their interconnections are recognised and defined. A common practice is to have a series of sub-networks representing the packages and linking these into a master network through several interface events, i.e. those which are common to both the master network and the sub-networks, along the lines indicated for the PP Ltd case in Fig. 48. This may at first give the impression that the calculations will become horribly complicated, but with computer packages it is possible to process each sub-network in the context of the master network, taking account of all the interactions.

MANAGING THE PROJECT ONCE IT HAS STARTED

Great emphasis has been laid on the importance of planning the project before it gets to the implementation stage where commitments are made and significant sums of money are expended. Project management is by no means confined to planning, and the control of the project once it really starts remains as the major function. Chapter 2 gave an introduction to the concept of control and the way in which it is exercised in projects, and this was followed by consideration of time, cost, and quality in Chapters 5, 6 and 7 respectively; this is now further expanded. Any control process will be based on the comparison at each stage of work of what has been done with what has been planned to be done, as illustrated by the plan—measure—control cycle, shown in Fig. 2. This will involve the collection of information on what has been achieved at each stage of review.

Control of time

It is often common practice to examine time progress at regular intervals throughout the project, but in many cases it may be more useful to examine key stages of the project, e.g. in the PP Ltd case examples of key stages or 'milestones' would be

- the decision on the product range (event 3)
- the granting of planning permission for the building (event 24)
- completion of product design stage II (event 9)
- the completion of building works together with the delivery of the production machinery (event 28).

Arbitrarily selected dates for progress meetings may be more convenient for those taking part, but it is more useful in control terms to pick a milestone date, tick off those items that have been completed and then examine in detail those which should

157

have been finished but are not yet complete. This does not mean however that control is left to the 'milestone meetings', as the project manager should be keeping a close eye on every activity, so as to get early warning of cases where work is slipping behind schedule. Certainly PP Ltd's project manager should know each week how every activity planned for that week (or earlier) is progressing, so that action can be taken where necessary.

The means of gathering information on progress will depend upon the nature and location of the work. For example, progress on the building of PP Ltd's factory can be physically examined by walking around the site, marking off on a copy of the schedule all those items recently completed and highlighting those which are behind schedule. This should not be an informal visual inspection but should be properly organised and recorded. Data can then be fed into the network and the repercussions of any deviations between plan and performance can be examined. If at any stage it appears that project delays are threatened then action should be taken not only to halt any further delay, but also to recover time lost if possible. This applies even if the activities concerned are not critical, since the use of float can be a matter of concern as it restricts freedom of action later in the project, especially in terms of the manipulation of resources. The collection of progress data on off-site activities is a little more difficult. Consider for example the design and manufacture of the production machinery for the PP Ltd factory; this is likely to be the responsibility of a company some distance from the site, and arrangements will have to be made to ensure that accurate progress data is obtained. A simple telephone call from the project manager to the supplier's office may not be good enough, as the supplier may informally give a rather rosy picture of progress. The PP Ltd project manager should always have available a current update of the sub-network covering the equipment design and manufacture, and be able to interrogate it if possible by direct access to the supplier's computer. If such access is not possible then the project manager should visit the supplier's works to physically examine what has been done to date.

Control of cost

In most projects this will be a matter of setting up a data collection system, probably linked to the progress control described above, to record both the value of work completed and also the cost of achieving it. In many cases it will be possible to work out in advance the value of each activity; in the case of building work this would be the priced bill of quantities which form part of the contract, in the case of the supply of equipment or computer systems it would be based on quotations for relevant items, and so on. When work is then recorded as being complete it is a simple matter to calculate the value of work completed to date. Calculation of the costs incurred may be fairly straightforward for major 'bought-in' items, but is much more tedious in the case of trying to control in detail the items of work completed. For example, if some part of the building work is being carried out on a 'cost-plus' basis because it cannot be properly defined in advance, it is necessary to record all labour and materials together with plant used and to allocate these costs to individual activities. The source of such data is often operatives' time-sheets, and invoices for materials and plant, but these are designed for other purposes and can lead to inaccuracies. In any one situation this subject should be considered in advance and appropriate systems worked out; books by Pilcher (1985) and Harris and McCaffer, (1977) give useful detail for building projects.

Control of quality

This has been clearly stated in Chapter 7 as an area that should always be pre-determined and is not thereafter negotiable. This means that it is somewhat different from the control of time or cost where it is recognised that either of these two may deviate from the plan and that despite attempts to correct deviations, overrun may have to be accepted as inevitable. In the case of quality, the correct approach is that a failure to meet the prescribed quality or performance levels should never be accepted, and that any deficiencies must be made good at whatever cost or time penalty. Once this principle is accepted the question of quality control becomes conceptually simple: the standards set are sacrosanct and must be achieved.

Control of quality then becomes a matter of two separate measures; first there must be a general ethos of the prevention of defective work by the provision of good planning, full supervision, appropriate training, sufficient and explicit information, good working conditions, and all other factors which are conducive to work of the quality required. These all apply at both the planning and execution stages of the project. The second quality measure may be referred to as 'compliance', namely the process of checking that the standards set are actually achieved. It is in this area that particular problems arise for project management due to the one-off nature of the work. In steady-state manufacturing it is possible to monitor carefully various parameters of quality, perhaps a dimension or other property, and by detecting very small shifts in values predict when a process is likely to produce defective goods. This early warning can prompt corrective action to be taken before the process gets out of tolerance. Unfortunately in project work each activity is likely to be different in detail from its predecessor or successor and no such early warning can be given. This puts a greater responsibility on operatives and supervisors to ensure that work is up to the required standard, but there still remains the need to carry out quality checking. Some carry-over is possible, for example in the materials supplied since they may be used in several different tasks and locations, but there is often a need for 100% inspection of the on-site work carried out. This is especially important where items are built into a project and it becomes increasingly difficult to correct faults which are found at a later stage. For example if mistakes are made at the design stage of a new machine which is then manufactured then the correction may be very expensive and time-consuming. Similarly if the foundations of a building are found to include sub-standard concrete it is expensive to remove and rebuild them at any time, but is extremely costly to do so once columns and walls have been built on them. In much project work the responsibility for the checking of quality is separated from that of the original work, and it is important to consider this when setting up the project organisation, a topic which is discussed more fully in Chapter 11.

159

Quality assurance

This is a phrase which has come to have a specific meaning in the area of quality management. It might be thought to imply the process of making sure that prescribed standards are met, but what is really meant is the information system that is designed to implement the procedures and methods for setting standards, the locus of decision making and the measurement of quality together with all the necessary documentation. It is very fully set out in ISO 9000 and various other standards which are referred to in Chapter 7.

Control means influence

For anyone to exert control over anything they must have the power to change it in some way; they may not have to make a change if all is well, but they must be in a position to do so if it is necessary. This has important implications for the organisation of project management, and this is dealt with in Chapter 11, but there is one point of immediate importance. This is the undisputable fact that only things in the future can be influenced, and therefore project control must concentrate attention on what still has to be done. It is natural and sometimes valuable to have a 'post mortem' examination of faults of the past in order to prevent their repetition, but it is not productive to have long discussions of responsibility and blame at the expense of devoting time to correction and prevention of faults in the future. The project manager must always be looking ahead to what has still to be done, perhaps leaving the accountants and lawyers to argue over who has to pay for past errors.

Updating and re-planning

Very few, if any, projects will work absolutely according to plan; in fact if it is claimed that a project has run exactly to plan this may itself indicate that something is wrong. Projects often run well and indeed many now run ahead of schedule, but of course being ahead of the planned programme in itself means that it has not run to schedule. What is important is that the process of planning and scheduling should be frequently updated so that it always represents the reality of what is being done and what can realistically be done in the future. This means not only keeping it marked up with work completed, but adjusting the logic where experience has presented problems of sequence of work, altering durations of future activities where recent performance indicates that early estimates were wrong, and taking account of changes in the availability of and need for resources. Project management is indeed dynamic, perhaps more so than some other forms of production management where the process can be allowed to run once it has been set in the right direction. For most projects there is no single 'right direction', as it has to set off simultaneously in several directions at once along paths of different and uncertain lengths, with unknown obstacles, paths converging and diverging at many points, and all finally coming together at the end, but never to be repeated.

NEGOTIATION

A project manager will frequently be in a position of having to conduct a negotiation, not necessarily in the formal sense of a negotiated contract, or agreements with union representatives, but in the everyday conduct of the business of managing a project. This might be a matter of agreeing a delivery date for equipment, a completion date for part of the project, the protection of works, an interpretation of specifications, and so on. The general objective of conducting a negotiation is to find a solution which meets your own requirements as far as possible, and at the same time is acceptable to the opposing party to the negotiation. There is an art to this which can be learned, and has been developed extensively in the world of selling (Karrass, 1974), but many of the principles are relevant to project management.

- Do not take up a strong stance which may lock you into a face-saving solution which does not really suit either side.
- Taking up positions and arguing over them is time-consuming, may lead to no agreement and is hence a waste of time.
- Being nice is not the answer — you may make good friends and have short negotiations, but you will not achieve wise agreements.
- Should the approach be hard or soft? If you are soft you may lose and not achieve a wise solution. If you take a hard approach your opponent may simply dig in, and no solution is found.

There is a better approach than the hard or soft attitude, and that is to structure the 'game'.

- Separate the people from the problem. Commitment to an agreement can lead to a cooperative spirit, which will help establish good working relations. Argument can lead to anger and frustration and bring out the worst in people. Try to understand the other person's point of view, although you do not have to accept it. Good communication is vital, and this means listening carefully as well as speaking clearly.
- Focus on interests and avoid positions. Try to find common interests, e.g. getting on with the work, and finishing the project. Try to find mutual gain. Explain your interests first, and then propose a solution, preferably one to which the other party can simply answer 'yes'.
- Think of a range of solution options, by generating as many ideas as possible. Assess the alternatives using objective criteria which should be first agreed, based wherever possible on established and published standards, e.g. British standards or ISO standards.
- Be prepared for failure, and protect yourself against an agreement which you should reject. What is the best outcome if you do not achieve an agreement, and what will you do? If you are in a weak position do not expect to have a mid-way solution. Think about how you will deal with an unscrupulous opponent.
- Overall be prepared, with good information, clear and realistic objectives and plans of how you wish to proceed.

161

10 People in project management

This chapter discusses people both as individuals and in groups or organisations, and makes special reference to teamwork and project teams. It is a well-worn truism to say that management is all about the control of some people by other people. There is of course a great deal more than that to the subject, for example the planning and use of equipment, but what is really meant is that most processes consist of actions undertaken by someone at the behest of someone else; a very large number of instructions are based on decisions taken by one or more people, and then conveyed to others for action, often through the medium of human communication. For some high-technology industries this may no longer be generally true since some processes are now computer controlled, not only in terms of maintaining pre-set values of various production parameters, but also in 'making production decisions' on the basis of data received. This stage has not really been reached in project management where much if not all of the planning and control of work has to be undertaken by engineers and other project managers, and the execution of work carried out by a wide range of staff. People are therefore very much at the core of most project management processes. There are many different groups of people to be considered, not only the project managers themselves, but the client, the designer, the financial expert, the quality controller, the supervisors and the skilled operatives who actually carry out the physical work of many projects. Consideration must also be given to the ways in which people work together, either in teams or on committees. A subject of very great concern is the necessary characteristics required of project management personnel, and the ways in which their skills can be learned and developed. Consider first some of the roles that have to be filled in any project.

PROJECT SPONSOR

In large projects the sponsor may seem to be a big organisation, perhaps a private company or a government department. It is often the case however that there is one person within that organisation who is the moving force, and whose confidence must be gained by the project team, i.e. the people who have to execute the work. Most really successful projects do have such a person who may be thought of as a 'project champion', although this is a title which is unlikely to be posted on anyone's office door. Manufacturers have for a long time talked about a 'product champion' as someone who is totally dedicated to the successful development and launch of a new product, and will not give up in the face of all the difficulties which may arise. Industrial history is full of cases of products which might have been a big success, but did not emerge because of the absence of a product champion. In the same way many development projects are not completed simply because there is no-one whose enthusiasm for them will overcome all the problems which arise. Once the development stage is over and a decision has been made to proceed with the project, it is likely that the project will continue through to completion, but the extent to which it meets its targets will depend not only on the project manager and the project team, but also on whether or not there is someone actively playing the role of project champion. There is no formal position or line of authority for the project champion who may come from within the client team, the design team or the procurement and construction team. The main value of a project champion is the enthusiasm that can be engendered within the whole project team, helping the members to work together as a cooperative group, rather than as a conflict of adversaries.

PROJECT LEADER

This is a title which is often given to one of the project team, but really means something a little different from that of project manager. It is perhaps appropriate in small projects where one person coordinates the work of a small part-time team, or as the leader of a small group which forms part of a larger project team. It is wrong however to think of the title project leader as being synonymous with project manager; the latter title has implications of authority, control and responsibility. Later in this chapter the characteristic termed 'leadership' is discussed in relation to the desired attributes of a project manager.

PROJECT COORDINATOR

Coordination means the bringing together of a wide range of information, checking for inconsistencies and possible conflicts, and presenting them to the appropriate persons for consideration, decision and action. The true coordinator therefore usually has a great deal of detailed work to do, but has little or no power over the project nor responsibility for anything other than the quality of the information handling and presentation. A project coordinator is a vital member of the project team which will depend upon the accuracy and timeliness of the information presented to it. Considerable skill therefore is needed to assess the quality of information received, to check it and then present it in a meaningful way to those who have to make decisions based upon it. Knowledge of computer systems will often be required, as well as a good understanding of the nature of the project work being managed. It is of importance to note that the characteristic of 'thoroughness' which is frequently referred to in this book is clearly of importance to a project coordinator who is handling data input to a network planning package; the omission of one item can invalidate the whole of a programme in the same way that the omission of one piece of machinery can frustrate the completion of a project. Thoroughness in project management is vital.

PROJECT MANAGER

The designation of project manager is one which has come into very wide use, and many would say that it is too loosely used. As was stated in an earlier chapter it is possible to see jobs advertised as 'project manager' which are anything from a volunteer to run a self-help group to the most important post in a multi-million pound undertaking. This section of the book seeks to set out what the responsibilities of a professional project manager are, and what knowledge, skills and characteristics are needed to perform the role. It is fairly simple to encapsulate the main responsibility of a project manager as 'hitting the project targets', namely achieving the completion of the project within time, at or under budget cost, and in accordance with the required performance and quality levels. It should not be necessary to add that all of this must be done not only within the law but within normally accepted good business practice. This statement is added here because there are some organisations which achieve their own project objectives, but in the process harm and/or alienate the firms with which they have been working. While this may produce good short-term results it is in the long term counter-productive: if a sub-contractor is forced into liquidation, this will reduce the number of competent firms able to carry out similar work in future, and so on. In a similar way if a project client adopts a very harsh line in the interpretation of contract terms, the professional project team may find mistrust on the part of suppliers and contractors in future projects for other clients.

How does the project manager manage?

This will depend upon the nature of the project and the position in which the project manager is placed. Therefore not all of the responsibilities listed here will apply in all cases, but should be considered within the wider project team which includes the client and/or sponsor. This list to a large extent follows the one at the beginning of Chapter 3 in response to the question 'What has to be managed in a project?', but in this case the items are discussed in terms of the way in which the project manager has to carry them out.

Scope

On the assumption that the project manager has been appointed at a very early stage, the first major responsibility will be to define the scope and content of the project. This will even have to precede the setting up of the project team, since it will not be known what will be required of the team until at least a general outline of the project has been defined. The project manager, possibly acting alone or with minimal assistance, should work with the client or sponsor in developing the brief for the project. In order to do this it is necessary to gain the confidence of the client organisation, and to be able to work with its senior staff. The project manager will have to be experienced in the type of project, and will have to be of sufficiently strong character to instruct the clients in their own responsibilities, and to gain their whole-hearted commitment to the execution of the project; remember that the client has to be managed as well as the designers, suppliers and contractors. Part of the project definition will be to determine the means of procurement, i.e. by the award of contracts, or the hiring of teams for design, supply and installation, or even in-house direct work. This will all have to be set out in the project manual by the project manager.

The project team

The next important task will be to set up the project team, which in the case of a substantial project may include some full-time project management experts, as well as part-time representatives of the various firms engaged in the work (Briner et al., 1990). In the case of a small project the role of manager may be carried out by one of the professionals engaged in the design or execution, or by a member of the client's staff.

Other team members should include staff with a knowledge of contracts, often quantity surveyors in building projects, or purchasing officers in plant acquisition. Information gatherers are needed on all projects, and this can be undertaken by a variety of people, but it is a good idea that they are independent of the providers of goods or services who may be tempted to express progress in an optimistic way. For example, the supplier of some sub-contracted goods may happily state that 'work is 50% complete' on a particular day. For a start '50%' sounds like a rough guess anyway, but does it mean that 50% of the expenditure on the work has actually been reached or simply committed? Does it mean that 50% of the estimated time has been spent usefully on the work or has half the time simply elapsed? What is really important is how much time is still required to finish the work, since that will be the best predictor of its completion and it is only work in the future which can be controlled. Much of the work of data collection is entrusted to junior members of the project team, and it is important that they understand the need to get good quality information and do not allow themselves to be hoodwinked by other people. This may in turn mean that an attitude of scepticism must be engendered in all project team members, and that checks are made on the quality of information to ensure that it is reliable; this is all part of the watchword of project management, namely 'thoroughness'.

Project planning and control

These are two of the main tasks of the project manager and team, and clearly it is important that these jobs are undertaken by suitably qualified and experienced people. Knowledge of appropriate planning methods is essential, almost certainly involving some form of network, to be used at least at the initial analytical stage, and probably throughout the duration of the work. These and other techniques of project management are then also used in the measurement of progress and the exertion of control of the work. In order to be able to have effective control it is important that the project manager has the power and authority to make changes to what is being done or the way in which it is being done. This is where a major difference arises between a project coordinator who simply assembles information for others to use, and a project manager who is empowered to actually influence work by executive action. The objectives of planning and control are to achieve the time, cost, performance and quality targets and all the methods described in Chapters 5, 6, 7 and 8 are available to assist in this. The subject of the control of quality may be regarded separately. On very large projects, such as defence installations, it may be the case that there is a whole team of specialists working on the quality management of the work, and this team should really come within the project management team, as it has already been shown that while quality is not negotiable it does have an impact on time and cost. Where there is no formal requirement for a distinct 'quality control office' which would normally be needed with a project under ISO 9000 etc., the responsibility for quality may be placed with the project manager. It is traditional in building contracts that quality is supervised by the architect, but this is usually a matter of accepting or rejecting the work which has been completed. Chapter 7 has already shown that there is much more to quality management than simply accepting or rejecting work, and that the concept of total quality seeks to prevent defective work from ever arising, rather than correcting it after the event. Hence there is a need for the project manager and team to have expertise in the subject of quality management, and to have the determination and sometimes even the courage to insist that standards are met.

The management of risk

Risk management is one of the tasks on the original list of Chapter 3, and the topic is discussed more fully in Chapter 13. Here it is sufficient to note that the one-off nature of projects makes them very prone to a range of risks, both chance events such as fire or flood, and the much more common risk of variation in time and cost compared with the plan. Project managers should know how to assess these risks and take action either to avoid them or to deal with them. Some of the more 'macho' project managers may be tempted to accept risks and hope that they will just never arise. This approach may work in cases of bullying suppliers into meeting schedules, but it is not possible to frighten off a flood or fire which can only be guarded against by physical prevention and/or by taking adequate insurance cover. Again it is the characteristic of thoroughness which will lead the project manager to take the appropriate preventive or protection measures to minimise the impact of risk.

People management

Project management is very much concerned with the interactions between people in terms of passing information, issuing instructions, controlling the work of other people, making changes, and educating and training the less experienced (Stallworthy and Kharbanda, 1990; Craig and Jassim,1995; Langford et al., 1995). The range of skills required is wide, and these have to be reinforced by personal attributes. There have been many attempts to define the characteristics of a good project manager, and certainly there is no universal agreement on what these should be. The heavy civil engineering industry has in the past employed well-experienced, hard-headed engineers and tradesmen as project managers, and put great trust in them, but it is now seen that another level of skills is necessary to manage a large project. Some time ago a senior contractor stated in a list of the characteristics important for a project manager, that the willingness to work very long hours was of prime importance. This does seem to be a little short-sighted, as what is of importance is what is achieved rather than how long it takes. Furthermore, if a project manager spends every waking hour on the project there is a very severe risk of overwork and stress which will possibly have bad consequences for the project. It is far better that the project manager sets up a good team and inspires them to work well, and then lets them get on with the work under his or her general supervision. The same speaker listed other attributes including an outward-going balanced personality, the ability to get on with people, good oral communication skills, and a willingness to take or accept responsibility. These are all appropriate and important, but while they may be seen as necessary they are by no means sufficient, and really apply to a manager in any situation whether it be manufacturing, selling or administration. In addition it is not enough to think only of oral skills when referring to communication, as the ability to write well is vital, and the ability to illustrate on paper or by computer can also be very valuable. There is also of course the skill of being a good receiver of communication, whether it be listening, or reading either the written word or the computer screen. There have been better attempts to list the necessary personal characteristics of project managers, which have included the following.

- The ability to look ahead and anticipate what should happen and what may happen, i.e. planning skills combined with foresight.
- Understanding measurement, ensuring that data collected is correct, relevant and adequate.
- Insisting on proper control, influencing what is to be done with determination.
- A good grasp of the one-off nature of projects, and the ways in which their management differs from that of other forms of activity. The concept of 'getting it right first time' is particularly important, as is the understanding of the inevitable sequencing of work. The extent to which resource limitations influence progress must be appreciated.
- An appreciation that the proper use of float is one of the ways in which projects can be brought in on time; it must be conserved and not allowed to be used indiscriminately.
- A reasonable working knowledge of the particular industrial environment of the project, namely both the technology or a part thereof, the commercial or contractual structure, and the general legal, economic and social environment.
- Realisation that nearly all of the personnel working on projects are supreme optimists when it comes to agreeing progress and delivery dates.
- The need to keep everything in proportion, and not to allow interesting trivia to occupy too much attention.

- Communication skills of writing and reading well, speaking and listening effectively, remembering that in management as in many activities communication must be a two-way process.
- The ability to have a simultaneous understanding of several problems, but not to get confused by this and to retain the ability to tackle and solve them individually.

If it is desired to condense this list into a few words it could be said that the project manager must have foresight, numeracy, control; reading, writing, speaking and listening skills; relevant experience and technical knowledge; self-confidence and single-mindedness combined with breadth of vision; and last but not least, thoroughness. In *Body of Knowledge* (1996), the APM has its own 'Personal profile of the Certificated Project Manager' which sets out the characteristics required of someone operating at a senior level in the profession.

To many people the above list would describe leadership (Jessen, 1992). This idea is worthy of further discussion, especially as many people see leadership as one of the most important qualities of any manager. The so-called 'born leader' is usually someone of vision, who has the drive to explore new areas and to enthuse others to follow. These qualities often lead into the unknown and carry a significant risk of failure, with the possibility of disastrous consequences. Many ventures cannot be achieved without entrepreneurial flair and the risk that goes with it, and it is in such ventures that a strong leader will succeed where the more cautious may fail. A powerful leader may not always lead in the right direction, as can be seen from many cases in politics and business, and may be primarily interested in the power wielded. Certainly an enthusiastic leader is not always interested in ensuring that everyone is on board and may leave stragglers behind.

This contrasts with what is required of a project manager (House, 1988) who usually has a target which has been fixed by someone else and the purpose of project management is to achieve that target, without omitting or losing anything at all along the way. The project manager must ensure that every little detail is completed as required — again it is a matter of being thorough. There is a case to be made for strong leadership in some projects, especially large prestigious or controversial ones, and this is where a project champion is essential. (See discussion earlier in this chapter.) This is important for the project sponsors who may have difficulties persuading the funding organisations to support the work. There may be public relations problems if it is perceived that the project raises objections from sections of the public or industry. In such cases the project champion may have to bring a great deal of persuasion to bear, and will have to generate wide support — namely great leadership is needed. The project champion alone will not however ensure the project is completed on time, at cost and to specification; it will need a project manager of the type described earlier to ensure success. It may be thought to be an ideal to have in one person someone who can at the same time show entrepreneurial leadership and the meticulous thoroughness needed to bring about completion in every detail. It is almost impossible to imagine such a person because some of the attributes of the two characters are in opposition to each other. What is probably the best approach is to have two people who by working in their respective roles can each bring out the best in the other, and ensure success for the project.

Stakeholders

At the time of writing this book the word 'stakeholder' has become very popular with politicians, and again we find a word which has specific use in a technical or business context being somewhat blurred in its meaning by popular usage. (The outstanding example of the change of use of a word in the project management field is 'Internet', which for 30 years was the name of the world conference organisation in project management, a major contributor to the development of the subject, but now faced with finding a new title because the word is universally understood to represent the global information network.) Within the project field the term stakeholder is used to mean any person or body of people who have a real interest in the execution or outcome of a project. In a typical building project there are clearly the following, who might be termed primary stakeholders

- the client company
- the architect, engineer and other construction professionals
- the banks and other financiers
- the local planning and building authorities
- the contractors and sub-contractors
- the suppliers of materials and components

(A broadly similar list will exist for non-construction projects, e.g. in the case of a production machine installation the only difference might be to replace the planning authority with the Health and Safety Executive.) In addition it can be said that there are many other groups who may be aware of the impact of the project, and while not formally party to its execution or ownership are directly affected by it, in financial or other ways; they may be referred to as secondary stakeholders.

- Organisations and people who will be users of the buildings, either as tenants, staff or general public. (For example, opera-goers have a very real interest in the general ambience of a new opera house, and may also consider themselves to be financial stakeholders, especially if it was funded out of the national lottery.)
- People who live or work near the project site and may be affected by it either during construction or in its subsequent operation.
- A project such as a new out-of-town shopping centre can have a dramatic effect upon small traders in the centre of towns, and this may in turn lead to a form of urban decay with all its undesirable consequences.
- Where a new project consists of a production facility it can enhance the employment within an area, and provide a significant boost to the local economy, with advantages to many local groups.

There are clearly many different parties and organisations which may be directly affected by the development of new projects over which they have little or no control. It is the duty of planning authorities to take account of many of these interests, and they certainly do this in relation to pollution, traffic and similar aspects.

It may however be relevant for project managers to give thought to other similar matters, especially where they can influence them or be influenced by them. For example, the extensive and well-organised protests which have been seen in recent years on the construction sites of new major road-works have presented to project managers a range of human relations problems that they previously never anticipated. For such project managers is it now necessary to have on their list of skills the ability to deal effectively with public protests without causing injury or damage; it is perhaps unfair to give project managers this responsibility which should be taken by local or national authorities.

Taking account of the stakeholders' interests   At the start of the conceptual phase of a new project it is important that the interests of all primary stakeholders are recognised. This will influence the way in which the project organisation is set up, i.e. the procurement route which is chosen. This is the detailed subject of Chapter 11. At the design stage it is possible to take account of many of the needs of both the primary and many of the secondary stakeholders. In the case of the design of an information system for a company with the general public as customers, the data handled must serve the needs of invoicing, stock control, accounting etc., but must also be easily handled and understood by customers. In an office-building project it is not only the owner and tenant whose interests must be satisfied, but also the customers who visit the building and the general public who have to walk by and look at it. Of course some cynics suggest that it is the last factor which predominates in the mind of some architects, but this could be the subject of a very long debate.

At a more detailed level it is the responsibility of designers to ensure that adequate and timely information is given by them to suppliers and contractors, with a specification of performance standards and the means of their control. It is up to the project manager to exert control over time, cost and quality, and during the implementation of a project all parties must be kept informed of progress, with all conflicts resolved. This will mean keeping the client and the financial bodies fully aware of all matters which concern them. It is in such ways that the interests of primary stakeholders are guarded. The interests of secondary stakeholders are somewhat less clear but are nonetheless important. Many of these should be taken into account at the project appraisal stage, which often takes place before the project manager is appointed. Project appraisal is the subject of Chapter 12 and is usually carried out by the client with external financial and technical advice, but often without adequate consideration of the project management aspects.

Obstacles faced by project managers

Like nearly all people with responsible jobs there are hazards to be faced by the project manager. Perhaps first among these is the fact that the project manager is usually highly visible, and there is no doubt who holds that position on each project. This is perhaps rather different from the branch manager of a manufacturing company or a retail branch where there is usually 'someone at head office' who can be blamed when things go wrong. The project manager on a building project is effectively in charge of a 'branch' of the company. One of the major differences however is that many more decisions are devolved to the manager of a project, and with that goes the responsibility for those decisions. Often materials will be locally bought, and labour locally recruited, possibly in an area where the local work environment is not well known to the project team. It may be not only the geographical environment that is new to the project team, but possibly also the type of work or the industrial background of the project. In some locations the project will not be welcome by at least one section of the community even if it is greeted by others. The project manager for a computer installation or the takeover and amalgamation of another company may be able to hide in a downtown office, but on a construction project everyone will know where the project manager's office is located. It is common practice for many project managers to take at least one walk around the site each day, partly to keep informed of progress, but really to show an interest and to demonstrate who is in charge; in doing this the project manager is deliberately seeking visibility and cannot complain about not being able to hide. Such a daily tour is more concerned with image than monitoring progress, which is a task which should be much more rigorously carried out by a junior member of the project team.

Given that there is an element of uncertainty about the progress of a project, due to its unique nature, the manager is likely to have to face many unexpected problems, with the consequent risks. This does mean that being a project manager is in itself a high-risk job. Later in this chapter there is a discussion of where the authority and responsibility of a project manager is based, but there is no doubt that many such people are placed in positions of responsibility without being given the authority to go with it; this can make life very difficult for the manager of any type of project.

Another situation arising in some projects is that the organisational structure is not fully worked out and the manager may find that work cuts across many boundaries, and these may change from time to time. Again the one-off nature of projects means that there is often no fixed set of rules within which the project operates. This can sometimes be a problem, but at other times may present opportunities which would not be open to someone working in a steady-state industry. Hence the project manager must sometimes be an opportunist and be prepared to act quickly when a window of opportunity presents itself. This is often a very public action, such as the launch of an off-shore platform when there is a lull in bad weather; everyone will be watching, perhaps even television cameras, success will be rather normal and non-dramatic, but failure may well be very spectacular.

Which way should the project manager be looking?

In the very interesting book *Project Leadership*, Briner et al. (1990) devote significant attention to the role of the individual and Fig. 55 is an extension of some of their thinking. In some ways the job of project manager is encapsulated by answering the above question. The diagram is drawn as a circle, not to indicate that the project manager must be looking in all directions at once for this would just cause dizziness; it is simply an indication that all directions must be considered, without any significant priority. It could be said that the prime responsibility is to the project sponsor, especially in formal terms if the sponsor employs the manager directly. In the true

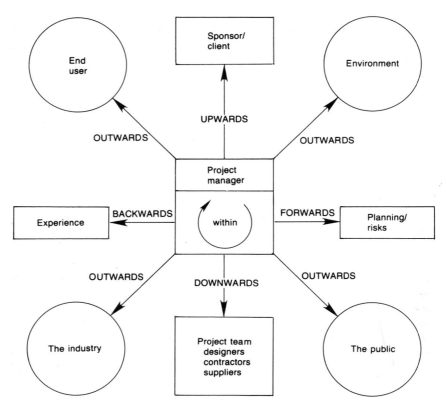

Fig. 55. Which way should the project manager be looking?

interest of good project management all directions are essential, although their relative importance may vary from one project to another. For example, good control of the project team is vital for the successful completion of the project. Taking each direction in turn, the project manager must look in the directions indicated below.

Upwards

This will usually be to the sponsor, and the prime objective is to deliver the project at the specified standard, on time and within budget. It will usually be necessary to manage the sponsor's input in terms of decisions, approvals and finance.

Forwards

There are two major aspects of looking forwards, first that which should happen, namely planning and scheduling, and secondly that which may happen, i.e. what may go wrong. The first of these has been dealt with in Chapters 5, 6 and 9, and the second is really the subject of risk management which is covered in Chapter 13.

Downwards

Here 'downwards' is in terms of organisational hierarchy, in that the project team is under the control of the project manager. It does not mean that all communications are downwards, and in fact most of the information flow in this link will be upwards, keeping the project manager fully informed of progress, and more specifically any ways in which actual progress has varied from that which was planned. Most of the downward flow will be decisions and instructions, as well as the less tangible aspects of motivation and encouragement. For the purposes of this discussion the term 'project team' is meant in the widest sense, and will include not only the various assistants to the project manager, but also the designers, contractors and suppliers, in fact any groups who are direct parties to the project.

Backwards

The important thing here is to learn from experience so that mistakes in the past are not repeated. It is not only mistakes that must be remembered, but also successes; for example the use of a particular form of contract as being very helpful, unusually good service from a sub-contractor, an innovative way of solving a technical problem, and so on. Luckily successes are easily remembered because they are looked back upon with pride; at the same time it does help to create a good project atmosphere if project successes are appropriately celebrated. On the other hand mistakes are often thought to be best forgotten, but it is important to learn the lessons of what lies behind the mistakes to ensure that they are not repeated. An example of this was a case where a large steel fabrication had to be cast into a major concrete pour on a power station project, and it was soon discovered that it had been placed 500 mm out of position. The costs of breaking it out, making good and then recasting were high both in terms of cash and time, and the project manager was understandably very angry. The reaction was to discipline the young engineer who had wrongly set out the work on site, and this certainly made all the other staff more careful about their work. This was not really enough because the system which had allowed one mistake had not really been changed, and it was possible that another similar mistake could arise. The real lesson which should have been learned was that the quality system was deficient, and a new system was needed incorporating an independent check.

There is one other aspect of looking backwards which should not be forgotten; it is wrong on any project to become preoccupied with the past, and wrong for project managers to spend their time 'fire-fighting'. The fact that this analogy is so frequently used implies that the approach is widespread in industry, but we hear relatively little about 'fire prevention' in this context. Perhaps this indicates that sorting out urgent and dramatic problems is more exciting and hence more glamorous than the quiet and deliberate avoidance of such occurrences. Many project meetings will discuss at length the cause of a problem that has arisen and who is responsible for it. It is important that the cause is identified so that a repetition can be avoided, but allocation of blame is of more interest to the accountants, surveyors and lawyers who should be left to sort it out and let the project team get on with the rest of the work. It has been stated earlier in this book but is worth repeating that the only work that the project team can influence is that work which still has to be done, namely the work from 'time now' through to the end of the project. Far too often there is too much time spent in arguing where the project has really got to and who is responsible for any delays. A good project management information system should clearly state the present position of a project, and the project team should concentrate its attention on the tasks ahead, especially if there is any lost time to be recovered, in which case some replanning and rescheduling must be done.

Outwards

The diagram of Fig. 55 shows four different outward directions in which the project manager must look. The one most directly linked to the project is the end user of the facility which is the subject of the project. In the case of a new production process it is clear that the actual plant operators must be considered, and also anyone who either supplies to that process or receives product from it. In the case of a building it is not only the client company which must be considered; in a development project of an office building it is important to consider not only the developer, but also the tenants, who may not be identified at the time of construction, and therefore it is necessary to incorporate flexibility of finishes, services etc. The other ways in which the project manager must look outwards are perhaps less direct but nonetheless are still important. There is the physical environment which may not only affect the work of the project, e.g. weather, but also may be affected by the project in terms of pollution by effluent, noise, fumes and waste. There is the social environment where the project will come into contact with the public at large, and may have to directly deal with individuals or groups. The industry in which the project operates must also be considered, as long-term good working relations with designers, contractors and suppliers will be important in future projects.

Briner et al. (1990) also suggest that the project manager should look inwards and be self-analytical and self-assessing to improve personal performance. To that might be added the idea that the project manager should be taking an occasional look 'over the shoulder' to see what else is happening, and who is doing what. Like those in any other senior position project managers will have an eye on their next appointment, which in project work usually means the next project rather than promotion.

Further thoughts on project teams

One of the advantages of project management is that the team for any one project can be created to suit that project, and is rarely inherited from a previous one. This means that the project manager can select the most appropriate people both in terms of the interests they represent and their individual skills. It is the range of skills available that lends strength to the concept of a project team, and each member can be recruited to perform a specific function. Furthermore the team can change in its membership as the project proceeds.

Each project team should have one named person in control, usually the project manager, although the actual title may differ, e.g. project leader (referred to early in this chapter). It would be normal practice for this person to be full time, as it would be for other members of the project team, including one person operating the project management information system (PMIS) together with progress coordinators whose responsibility it is to seek information on how the project is proceeding. A quality manager may well be in the team full time, as may accountants or surveyors in control of cash flow and interim payments to contractors and suppliers. In addition to the full-time members of the team there will be several part-time members depending upon the nature of the project. Part time may mean either someone who has responsibilities in other projects or other situations, but may also include members who have only a short-term interest at one stage of a project. For example a lawyer may sit in on project meetings from time to time to advise on specific legal matters which may arise. The same lawyer may have a short-term full-time involvement if the project is at the stage of a public enquiry. It is likely that on a major construction project the main contractor will have a full-time member of the project team (possibly as the project manager!). Each sub-contractor and specialist supplier may have a part-time member at the appropriate stages, e.g. design, planning, and at the time of execution of their own work.

Where does the project manager fit within the organisation?

This question is the same as asking where the project management responsibility lies within the organisational structure of the project, and there are many answers to that question, which is dealt with in Chapter 11. For the moment however consider only the project manager in person and the full-time members of the project team.

The project manager could be drawn from the staff of any one of the following parties

(*a*) the client company
(*b*) the professional design team
(*c*) the main contractor
(*d*) a specialist project management consultancy firm
(*e*) a freelance individual.

The organisational base of the project manager is clearly important in terms of responsibility, but the professional background can be from any one of the above. As stated earlier in this chapter it is however of importance that the individual project manager does have experience of the appropriate industry in which the project is located. While some projects are managed without such direct experience there is no doubt that it is preferable, and certainly lends credibility to the person in control. The construction industry has drawn many of its project managers from the professions of engineering (whether as contractors or consultants), quantity surveying, and more recently architecture. In earlier centuries there was little division of responsibility between design and construction, and many of the great projects of the past were project managed by the architect or engineer. The quantity surveying profession came into being after this, and because of their work in measurement and payment it was a natural step for them to move towards project management, which they have done successfully. There are other professions involved in some projects, especially the specialist consultancies who may draw their staff from almost any background before training them in the approaches and methods of project management.

In projects within other industries, for example the re-equipment of a factory, the project manager may well be an engineer or production manager on the staff of the client company, or that of the company designing and/or supplying the new equipment. It could of course be that an independent project manager will be appointed. The computer industry has itself become somewhat specialised in the practice of project management and is likely to offer this as part of the design of new systems and their installation.

What is clearly emerging is the recognition of project management as a separate discipline, and the need to educate and train people specifically for this type of work. It is no longer good enough to send staff on a short course and expect them to emerge as competent project managers. There are now both undergraduate and postgraduate courses in the subject, and professional recognition through the APM. As was emphasised earlier, education alone is not enough, and appropriate experience is essential. The APM takes account of this in its membership criteria, and the award of the title Certificated Project Manager, which is now well recognised. This is important because the simple title of 'project manager' is not protected, and is often given to someone in charge of a small operation which has few of the characteristics of a complex one-off undertaking.

For discussion of the authority and responsibility of the project manager it is necessary to look at the organisational structure of projects. This will be examined in Chapter 11. See also Appendix 3, *Qualifications in project management*.

Typical project team for the Photo Products Ltd case study

The following is not intended as a model to be used in setting up any project team, but can be regarded simply as an example of what might be done in one particular case. Every project will be different and the team structure and its responsibilities will have to be worked out in each case; indeed it is one of the first tasks in any project to determine how it will be managed and by whom.

The PP Ltd case is taken up at the point where a decision has been taken to proceed with the new venture of introducing a new product, a camera, which is a departure from the company's previous range of materials and other supplies to the photographic industry. The company therefore has a good understanding of the industry but is setting up a totally new facility for the new product. The senior management decide to manage the overall project themselves and release one of their existing production managers to undertake the task, with the intention that the same person should ultimately become the general manager of the new facility. It is recognised that at various stages other specialists will become involved in the project management team, and that appropriate responsibilities will be devolved. Since this is a major development for the company it is further decided that the project manager should report directly to the chief executive of PP Ltd, who should in turn report regularly on the project to the main board.

The PP Ltd project team is then made up of the following.

- The project manager, in overall charge of the project, reporting to the chief executive; long experience of the company and the industry, but relatively little experience of project management. (Note that this last point is a reflection of what happens in practice rather than what is desirable.) It is a full-time, fixed-term appointment. It would be important for the person selected to be given appropriate training including experience of similar undertakings. One advantage of the project manager being recruited from the client's own staff is that it enhances the involvement and commitment of the client to the project. There are many instances where a client has held up progress without being fully aware of the fact, and it has been stated several times in this book that the client has also to be managed in its dealings with the project. Not only will the PP Ltd project manager be able to overcome some of these problems, but will also be able to carry over much of the project information into the long-term running of the new plant, both in terms of factual data and the reasons why certain actions were taken.

- An assistant project manager, someone with extensive experience of the management of projects of this nature, namely the setting up of new production facilities. Recruited from outside the company to bring into the team the necessary skills in project management. Again a full-time appointment.

- A project information officer, or similar title, would be someone with the appropriate skills to operate a computerised project management information system. Such a person would also be responsible for the gathering of information, and must therefore be competent to judge the quality of that information, and to take appropriate action to ensure that it is obtained. This might commence as a part-time responsibility, but once the project is in full swing it would become full time.

- An estate consultant would be engaged to carry out the search for a site, and then its acquisition. A part-time member of the team, mostly at the start of the project.

- A product design specialist, possibly a permanent appointment to the company as the nucleus of a design team, but not a full-time member of the project team. Alternatively the design of the new product might be carried out by a consultant specialist.

- A production specialist, whose main responsibility is to design the manufacturing system, and supervise its installation and commissioning. PP Ltd have already decided that this person should be a newly recruited engineer who will thereafter become the works manager for the new plant. This is one of the key posts in the project team and should be a full-time one.

- The responsibility for the management of the building design and construction could be handled in a variety of ways which are discussed in Chapter 11. PP Ltd have concluded that they do not wish to have detailed responsibility for this part of the work and have appointed an architect as leader of the design team, coordinating all aspects of the construction. The form of building contract to be used is based on a fixed price and completion date, but with provision for a limited amount of addition or variation to the work. It is therefore appropriate for the architect or nominee to be a part-time member of the project management team. It is also likely that the contractors would appoint their own project manager to be in charge of the on-site construction, and such a person would also be a part-time member of the client's project team.

- Responsibility for the marketing of the new product will be given to the existing marketing department, which will therefore be part of the project team, at least in the early and late stages. Similarly there will need to be an input from the PP Ltd. administration to set up the appropriate financial and personnel procedures etc. The buying department will be entering new areas and should also be represented on a part-time basis at the appropriate stages.

There may be other groups which would be represented at times on the project team, perhaps the funding organisation which may want to be assured that the project runs smoothly, the local planning authority, employment agencies, and so on, all having an interest at various stages.

Managing the team Quite apart from managing the actual project, the team itself has to be led and managed effectively and it is in this respect that the project manager will have to bring to bear the skills of leadership, tact and persuasion. There are very good texts available on the management of teams, including the one by Briner et al. (1990), and this is well worth studying for anyone who has to set up a project team such as PP Ltd in the case described above. There are various factors to be considered.

As is often the case the PP Ltd team membership will not be homogeneous, and will include not only a range of skills, but also a range of backgrounds, attitudes and also loyalties. It is likely that all the full-time members of the team will be loyal to the project manager and the client, but some of the part-time members will at best have a divided loyalty, partly to the team and client, but at the same time to their own employer, e.g. the building contractor will command the loyalty of its own manager on site. In many cases this will coincide with the best interests of the client; however in times of direct conflict it is not possible for both loyalties to pull in the same direction. The mixture of full-time and part-time members with their different loyalties does have an impact on the management style that the project manager can use. A commanding style can be used in the case of subordinates in the team, e.g. the assistant project manager, but in the case of representatives from other companies the style must be more persuasive, using the carrot rather than the stick. The project manager must be adept at using the most appropriate methods at the right times.

Good communication This is very important, not only to ensure that no information is missing or wrong, but also to create a spirit of co-operation among the members — the so-called team spirit — and the feeling of having a common objective, namely the successful completion of the project to everyone's satisfaction. In every project there will be good times and bad times. There is every prospect that mistakes or shortcomings will be given a great deal of attention with special 'post-mortem' sessions to find causes and attach blame. At the same time it must be remembered that success should be celebrated, as this will do more to enhance performance in the future. An interesting case is often quoted where a large party was held to celebrate a project, but it was held before rather than after the project was complete; it actually served a very useful purpose in engendering a team spirit which proved useful in the subsequent work. If it had been held at the time of project completion, that goodwill would largely be wasted as there was no more work to do and the team were on the point of disbanding.

As has been stated earlier in this book good communication is not only a matter of being able to get your own ideas across, but also being sure that you receive the right messages from the people who are speaking to you; this is usually a matter of being a good listener. Generally we tend to hear what we want to hear, and ignore the rest. There are some good tips about how to improve the efficiency of listening.

(a) Reduce the number of links in the chain (think of the game of Chinese whispers).
(b) Do not let your mind wander on to what you are going to say, concentrate on listening.
(c) Listen to concepts rather than the style of presentation.
(d) Do not get agitated and do not jump to conclusions.

Project progress meetings   These often have a regular place in the diary of all the team members, but this is not really the best way to manage a project. First, meetings should only be called if they are necessary and if there is a specific agenda to discuss. Only those members required for that agenda should be called; if people feel that their time has been wasted on an irrelevant meeting they may decide not to attend the next one, just when their presence is essential. While the idea of a regular monthly or weekly meeting is attractive in that it makes diaries easier, it often means that meetings are held at the wrong time. From a project point of view the right time is at significant key dates in the project; at such times it is possible to check whether critical jobs are actually complete or not. For a regular meeting to be told that the next key date will be met 'in about a week' is not good enough.

The project team should not only examine the progress of the work it is managing, but should also look at its own effectiveness in doing so. How good is it at taking decisions and then implementing them? Is its information up-to-date and accurate? Perhaps most importantly is the project being properly managed in terms of cost, time and quality?

MANAGEMENT DEVELOPMENT — THE TRAINING OF PROJECT MANAGERS

At present project management is regarded as a secondary or higher level skill which is learned by those who are already trained in engineering, computing, architecture, building, production, systems or other disciplines. Project management is not seen by many as a primary qualification or the subject of a first degree, but there are many postgraduate courses in the subject. Most of the masters' courses lead to an MSc, but there is at least one MPM (Master of Project Management), and now most MBA programmes include project management, at least as an option. Most of the postgraduate courses which are devoted largely to project management are comprehensive, and include a wide range of topics, together with an intellectual analysis of how the discipline has evolved and functions.

At the level of day-to-day work as a project manager there are several aspects of training which are very important. As is the case in many subjects there are managers who take the view that the only way to learn is by being 'thrown in at the deep end', but this carries the risk of drowning, and taking other people and the project down at the same time. Courses alone certainly cannot provide a full education in any subject, and therefore a blend of experience and formal tuition is perhaps the best route. For those currently engaged in project-based industry there are many short courses, ranging from one-day introductions to others which run over several weeks, often in conjunction with practical experience.

The APM has acknowledged the need for this blend of experience and formal learning, and has recognised this in its formal qualifications. At the first level of qualification is the APM Professional (APMP), which is assessed by a two-part examination which tests individuals' understanding of project management, and their approach to it. After this, full membership of APM is based on experience, and then a higher stage of Certificated Project Manager can be achieved by success in a comprehensive profile assessment, which requires considerable knowledge, experience and competence. Some Members with appropriate achievement may be elected to Fellowship. Both the APMP and the Certificated Project Manager qualifications are based on the *Body of Knowledge* (APM, 1996) which is now recognised as the basic European standard for project management competence.

There are now also National Vocational Qualification (NVQs) specifically relating to project management in construction (see Appendix 3).

PROFESSIONAL BODIES IN PROJECT MANAGEMENT

In the UK the APM was established in 1972, and has over the following quarter of a century developed into a leading professional body, with recognised qualifications. After some deliberation it voted in 1996 on a proposal to change its name to The Association for Project Management, to reflect its developing role. The Association has its headquarters in High Wycombe, but has a network of local groups throughout the UK. The association is strongly linked through the International Project Management Association to many other national project management organisations, mostly in Europe. There is another major professional body, based in USA, The Project Management Institute which also has well recognised routes to qualification. Many other national groups are linked to either the APM or PMI.

11 Procurement

Procurement embraces the acquisition of services, goods and materials which together make up a project. Chapter 9 which was concerned with project implementation listed most of the steps involved, but said relatively little about how this should be achieved. There is no single route to the procurement of a facility, and the range of approaches is very wide, depending upon the nature of the project, its size, and the capability and wishes of the client and/or user.

METHODS OF PROCUREMENT

One of the early decisions to be taken by the project manager will be which method of procurement to follow. Some very simple examples may illustrate this point. At one extreme, a client wishing to set up in a particular business may simply seek to buy an existing company and take over everything that goes with it, including premises, staff, production plant, work in progress, customers and management systems. In this instance there may at first seem little to be done in the way of project management, but it does inevitably involve change, and this has to be managed. There will certainly be legal and financial steps to be taken, with valuations to be made of stocks of materials and finished goods; existing contractual arrangements with clients and suppliers must also be checked. Such a project does however offer some advantages of simplification; the buildings, plant and stocks can be actually seen; the time-scale can be very short; the staff are already in place, trained and experienced; overall the client has none of the uncertainties of other ways of getting started and can see exactly what is being acquired. There is effectively little risk in the actual process of procurement if reasonable care is taken; the final price is usually known and everything can be inspected before any commitment to pay is made. There remains of course the commercial risk of whether the business bought is worth the price paid, and that is more a matter of financial project appraisal, which is common to many projects and is discussed in Chapter 12. At the same time there may be disadvantages to this very simple method of buying an existing facility; it may not be quite what was wanted in terms of capacity, location, quality of product and efficiency, and as a result may be second best compared with a purpose-built plant.

At the other end of the spectrum is the type of project where the client company feels its way through the project step by step, buying or making the component parts as it proceeds. This approach may be taken in the so-called 'open' projects in which little is known of the final form of the project until the project itself is some way down the line; this is typified by development projects for new products and new processes, many of which are abandoned before they are very advanced. The application of project management to development projects is rather a special area of the subject and is dealt with in a number of texts, many of them relating to the chemical industry; many of the aspects of this work run counter to the generally accepted principles of project management. For example there may be no specific objective in view, there is often no target time-scale or even cost, some sequences of jobs may be recycled several times, there may be branching 'either/or' nodes or events (compared with usual networks where every task must be completed), and many projects are abandoned at an early stage. There is a distinct approach to such networks and this is given elementary consideration in Chapter 13 which deals with a number of aspects of uncertainty, development projects being one of them.

This does not mean that there are no determinate projects which are handled in the step by step way, sometimes referred to as 'hand-to-mouth', or 'do-it-yourself'. This does not refer to the domestic DIY enthusiast but to the client which has within its own organisation the range of skills required to carry out all the tasks of the project. Clearly a large building contractor would be able to take on the organisation and construction of a new office building for its own use, but there are many other examples. A general engineering company could directly buy and install new equipment using its own staff, a computer software house can design and implement its own accounting system, and so on. It is a very common approach used by large clients undertaking small projects — they are in a position of power over suppliers, have good management skills, and access to funding. For the purpose of this chapter this method of procurement is referred to as 'internal contracting'. The advantages of internal contracting include the following.

(a) The whole process of selecting and briefing other organisations is eliminated, with consequent savings of time and money.

(b) The step-by-step process allows decisions to be changed as the project proceeds.

(c) Where confidentiality is important this is helped by keeping all information within the client organisation.

(d) Actual work done may be carried out by existing staff who are not fully occupied, e.g. maintenance fitters are often needed to be on stand-by and can easily take on extra work.

(e) Quality standards are easier to maintain when work is carried out in-house by people who are familiar with the standards used by the client.

There are of course disadvantages of carrying out the work of a project in-house by internal contracting. It is assumed that by using the phrase 'internal contracting' that all work is properly costed and paid for, and it is not just a matter of the works manager ordering goods and charging them to a general account. There does remain the problem however that not all work will be properly estimated in advance of the start of the project (there may be no budget as such), and hence there is some uncertainty about the project cost and whether or not the cost of work is being contained. In a similar way if the work of the project is not properly planned in the first place, there is no schedule against which progress can be measured. Some aspects of the work might be better carried out by an outside specialist company, rather than internal staff who do not have the most appropriate skills or may have other demands on their time. There is of course no reason why some parts of the project should not be contracted out simply because other parts are being handled internally, but this might be an inefficient way of using outside contractors.

In all projects there are risks, even though steps may be taken to minimise them. Where a project is managed through internal contracting the whole of the project risk is borne by the client, simply because there is no-one else to carry the risk. This type of procurement is therefore at the opposite end of the spectrum from the one previously described where the client simply takes over an existing facility. Clearly there are methods of procurement which are intermediate between these two extremes, and these are now examined in some detail, with particular reference to the construction industry. There are analogous methods in use in other industries and these are referred to later, in connection with a review of the earlier case study of Photo Products Ltd.

BROAD TYPES OF CONTRACT IN CONSTRUCTION

Reference is deliberately made here to 'types' of contract rather than the 'form of contract' which really means a very specific and fully defined set of terms and conditions. Some discussion of these particular documents is included towards the end of this chapter. For the purpose of this present discussion six broad types of contract for the construction of building and related works are considered. They are listed in sequence of increasing involvement of the client and hence of increasing risk borne by the client. For the other party, the contractor, there is the complementary decline in risk, as is indicated in Fig. 56.

Turnkey contract

Here the contractor, or often a contracting consortium, will undertake to provide a total service including both design and construction, of a facility which will meet a specified set of performance criteria. This type of contract has often been used in the construction of chemical plants or power stations, but is now also being seen in other construction types. It does offer a number of advantages to the client.

- There is only one organisation with which the client has to deal.
- The contractor is able to bring to the design team the experience of site construction which helps achieve economy of design.
- The client has no responsibility in the design process, and simply defines the final output performance of the completed project works.

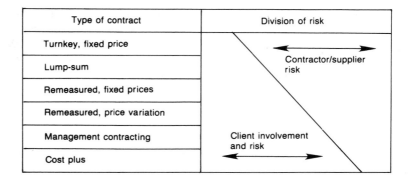

Fig. 56. Division of risk for different types of contract

- The integration of the design and construction phases within one responsibility will usually shorten the total project duration. This is often referred to as a 'fast-track' type of contract.
- The total price for the works is fixed and known in advance, with the important proviso that the client does not have a change of mind about the required performance, or want in any way to modify what is being built.
- The turnkey contract therefore has many of the advantages of buying an existing facility, but also achieves a new plant which is likely to be based on the latest technology available. It also is guaranteed to operate at the specified performance levels.

The disadvantages of turnkey contracts are relatively few in number but may still be important; the client may regret not having an input to the design and construction, but this is relatively minor. The major drawback seen by some clients is that they are asked to pay dearly for the advantage of having nearly all of the risks taken away from them and covered by the contractor, some indeed feel that it is an expensive way to procure a facility.

Lump-sum contract

In this type of contract it is usual for the design to be undertaken by a consultant architect or engineer, and for the work then to be fully defined in some way, either in terms of drawings or bills of quantities, or both. A contract is then placed with a contractor on the basis of a fixed total price. This gives the client considerable protection from risk which has to be borne largely by the contractor. It is however important that the work is not changed in scope or detail by either the client or the designer since this effectively cancels the concept of a lump sum in exchange for a fully pre-defined work package. This type of contract may have the complication of a fixed completion date built into it, with pre-determined penalties for late completion. Such penalties are often difficult to enforce, but compensation for delay may be sought in terms of liquidated damages, for which actual losses by the client have to be proved.

A lump-sum contract isolates the client from a number of the risks inherent in other contract types, but it does have the disadvantage of being somewhat inflexible. It is therefore not often used in large projects, but is fairly popular with clients offering small and short-term contracts. It has been seen by some as a way of avoiding legitimate claims by contractors, but this cannot always be guaranteed.

Remeasured contract with fixed prices

In this type of contract the work is specified by drawings and a bill of quantities, and the price for the work is determined by quoting both a quantity and a rate for every item in the bill, which may run to several hundreds or thousands (e.g. $£x$ per cubic metre for casting concrete in beams of a certain size). This means that the contract price is the sum of all these items, and this takes precedence over a total amount quoted at the end of the document if there is a difference between the two figures. In the event that changes are made to the scope of work during the contract then the work is 'remeasured', and priced at the rates in the bill so that the sum to be paid reflects exactly the work actually done. In a similar way if any work is deleted then its value is also deleted. There are sometimes clauses included to the effect that if the sum of changes is relatively large then other adjustments to the contract sum can be made. All the prices are quoted on a fixed basis, so that no claim can be made for changes in contractor's costs.

This contract type gives protection to the client in terms of escalation of material and labour costs, but at the same time means that changes in scope, whether at the client's request or because of unforeseen events (e.g. finding unrecorded pipes or cables which then have to be protected) do have to be paid for. In this way the client has to carry the risk of such changes, but since this relieves the contractor of them the contract price should be keener. It therefore provides a reasonable balance of risk, especially where there is some prospect of unknown factors arising. At the same time it allows clients to have the freedom to make changes at their own expense, and relieves them of having to commit themselves to early decisions.

Remeasured contract with price variation

This type of contract is similar to the previous one except that prices are not absolutely fixed. It is usual to fix them for a limited period, say one or two years, but thereafter change them by reference to nationally published figures. This type of contract is favoured at times of high inflation when contractors are reluctant to agree to fixed prices because of the risk that this involves. If forced to quote fixed prices their reaction would be to quote very high rates to guard against inflation. It does of course mean that clients have to bear the risk that prices may rise, but they are able to get more realistic bids in the first place, and contractors are protected from high inflation in the medium term. The work is remeasured as in the previous case, so that whatever work is actually done is paid for at the appropriate rate.

Management contracting and construction management

These two phrases are not synonymous but do have some features in common (Barrie and Poulson, 1992; Cushman et al., 1983). These methods of procurement have evolved in recent years, following to a great extent some of the best practices of project management. The essential feature of this approach is to make use of, and integrate into the design and construction, the best aspects of construction technology and its various applications. In the USA the method is referred to as 'construction management', and usually involves the appointment of a construction specialist as a professional adviser to the project team, often as a member of the design team along with the architect and engineers. In the UK the practice has been rather different, with a major contracting company acting as 'management contractor' in many projects. One of the main advantages of these approaches to contracts is to improve the coordination and overlap of design and construction, and generally to improve the management of design. Shorter project times have been achieved, to the advantage of both client and contractor, and this aspect has led to the use of the phrase 'fast-track'. Unlike the professional adviser, the management contractor has an involvement in the main contract, and is responsible for obtaining tenders for the work and awarding contracts for the work on behalf of the client. The management contractor is usually paid a fixed fee for providing this service (i.e. not a fixed percentage of the cost of work done), but the fee can often be varied if the time and or cost performance of the project differs significantly from the initial plan and budget. The managing contractor has a wide range of responsibilities, but also usually has the necessary skills to carry them out.

- Overall project management, including the preparation of the initial cost and time plans, and later the coordination of all work on site.
- Obtaining tenders for work and/or negotiating contracts, which are then placed with appropriate contractors and suppliers.
- Quality management of the project.
- Issuing payment certificates.
- The design of the works which may in some cases be undertaken by the managing contractor, or alternatively placed with a design firm.

It can therefore be seen that this method of contracting is basically one of 'cost plus', but has been considerably refined to avoid the drawbacks of that very simple approach which is described in the following section. Management contracting does ensure that the whole project is well managed, right from the initial design stage through to completion, using the skilled resources of a large organisation. The fee paid to the managing contractor is geared in such a way that there is a strong incentive to complete the project on time and within budget. It does still mean that the client is free to make changes to the scope of the work, at an appropriate price, and will have to meet any genuine costs which might arise during the course of the work. At the same time contractors are not asked to bear onerous risks for which they would normally expect to be paid higher basic rates. The use of management contracting has become popular, especially in larger projects, perhaps because it does take advantage of project management methods, and spreads risks to those who are most easily able to carry them.

Cost-plus contracts

This fairly simple method is based on the concept that a contractor is given instructions to carry out work as the project proceeds, sometimes with only rough prior notice of what is to be done. Because of the lack of information it is not possible to tender for such work other than by supplying a list of rates for each type of material, trade and plant. It is then usual practice for work to be authorised and paid for at these stated rates plus a fixed percentage to cover overheads and profit. There is a major drawback to this type of contract since there is no overall budget limit or even proper control, and there is an incentive through the plus element to increase the cost of the work. In some types of work, such as emergency repairs there may be few alternatives, and the practice is now restricted to such activities.

FORMS OF CONTRACT

This phrase is used to signify that a particular structure of contract documentation has been written to act as the basis of a contract between client and supplier or contractor, usually specific to each industry. In theory any one specific form of contract will be clearly written within the framework of any one of the broad types listed in the section above; in practice most of the widely used forms of contract will be based on one of the listed types which makes provision for remeasurement of work complete. Many of the forms in general use have been prepared by one or more of the professional institutions relevant to the work, and some of these are listed below.

The purpose of a detailed contract is to make quite clear what the work comprises, the way it should be measured and hence valued, and how and when the work should be paid for. Considerable attention is usually paid to sub-contractors and suppliers, and who is responsible for their work. A great deal of legal significance is attached to the contract, and hence it is useful to have a standard form which everyone will understand. However the nature of the work done in a project will vary widely in different industries, and it has been found necessary to have a form which is dedicated to the industry concerned. In the past some of the forms of contract have been prepared very much from the point of view of the client, and the contractor has to accept them and to work within them. All that a contractor can do is to take advantage of every opportunity that arises to gain benefit from the contract, and it is because of this that the culture of submitting claims has grown to the point where in some projects it seems to take precedence over doing a good job. More recent forms of contract are written with the objective of achieving a fair outcome for all parties to the contract, which means that all parties can concentrate their attention on getting on with the work. The comments below only give a general description of different forms of contract; the subject is clearly a complex one and any project manager should take appropriate specialist advice before deciding on which form to use on each project.

Civil Engineering

The so-called *ICE Conditions of Contract* document was prepared jointly by the Institution of Civil Engineers, the Association of Consulting Engineers, and the Federation of Civil Engineering Contractors, and has been in use for many years. Also in use is the *Civil Engineering Standard Method of Measurement* (CESMM), which sets out in great detail the way in which the work of a project should be specified and quantified. A good detailed guide to this is *The CESMM3 Handbook* (1991) by Martin Barnes, a Past Chairman of the APM. One of the significant aspects of CESMM is the introduction of 'method-related charges' which are intended to cover the cost of items which are not proportional to the completed measured works; this means that the measured items in the bill can be restricted to those costs which are directly proportional to the work completed. Tenderers are then given the opportunity to write in such method-related charges to cover the cost of work which does not appear in the bill of permanent works, e.g the provision of temporary works. This means that there is no basis for dispute if the scope and/or quantities of work are varied during construction. In relation to contracting for engineering construction a comprehensive but fairly simple book is written by Marsh (1984).

New Engineering Contract

In recent years a complete re-think of contract documentation in engineering has led to the publication by the Institution of Civil Engineers of the *New Engineering Contract* (1993). This is a very comprehensive approach to construction contracts, based firmly on the principles of project management, and formally recognising the role of 'The Project Manager'. It emanated from a wish to build on the Latham Report which promoted the concepts of trust and cooperation as replacement for the adversarial client/contractor approach. Early reports on the use of this contract form are promising, but until it has been subjected to a period of 'testing' in the courts of law, it will no doubt be seen by some parties as untried and unproved. If, as is hoped, it removes many of the causes for argument then it will indeed be a great step towards the more efficient management of projects. In the past there has been far too much time, money, and effort spent on arguing about contracts at the expense of getting on with a good job and giving all-round value for money. There have been criticisms of the new contract; many of these have come from lawyers who in the past have been heavily involved in contract work, and see possible loopholes in the new document.

Building construction

In a rather similar way to the ICE conditions discussed above, there is a well-known building contract form referred to as JCT80, which has evolved from an earlier form dating back about 80 years (Parris, 1985; Clamp, 1984). It was prepared by the Joint Contracts Tribunal, a body including many organisations representative of architects, surveyors, contractors, sub-contractors and public sector clients. It sets out clearly all the aspects of the contract and makes use of bills of quantities usually prepared by an independent quantity surveyor, and it makes provision for remeasurement and escalation of prices. As in the case of civil engineering the detail of this contract form is considerable, and anyone who needs to consider it should either consult an expert or refer to the useful book by John Parris.

One of the differences between civil engineering and building work relates to the way in which bills of quantities are handled. General practice has been that the bills of quantities in civil engineering are prepared by the consulting engineer in a project, but in a few cases this task has been handed on to a quantity surveyor. This contrasts with building projects where a quantity surveyor is nearly always responsible for this part of the consulting work. In building the contractor will rely on the bill to give a full description of the work to be completed, whereas in engineering the tenderer will rely more on the drawings and specification. This is partly because the engineer is responsible for the design of the works and the preparation of the bills, the budget estimating, the supervision and measurement of the work on site, and for the payment to the contractors. On most building projects these tasks are divided between the architect who is principally concerned with the design, and the quantity surveyor who is primarily concerned with financial matters. This separation is partly the reason for engineers becoming project managers on heavy construction, while in building the quantity surveyor has taken on this role. It may be for this same reason that project management has really flourished in engineering, but the division of responsibility in building has made this more difficult. Other differences between the two sides of the construction industry include the number and range of sub-contractors. In building work they are often suppliers of special parts of the work amounting to a high proportion of the total; indeed once the building frame is completed most of the work will be that of sub-contractors. By contrast the sub-contracted work on an engineering project usually consists of providing a specialised service with dedicated equipment, such as bored piles, and bulk earth-moving.

Property development

In 1984 'System for building design and construction' was published by the British Property Federation (BPF), a body which represents the property development industry. The Federation had concluded that their members could obtain an improved service from the construction industry, and commissioned the work from a specialist group which took into account the bringing together of design and construction under a project management umbrella. The preface to the BPF manual (BPF, 1983) says

> 'Members have become increasingly concerned about problems in building which occur far too frequently, particularly those of poor design, inadequate supervision and insufficient choice of material. Many contract methods cause delays, by their nature are inefficient, and can substantially increase costs. Worse, they cause and sustain "them and us" attitudes between consultants and contractors.'

The system is based on a number of concepts which are designed to improve overall performance of the project, including the following. At the outset a client's representative is appointed, with authority to issue instructions to the design consultants and contractors, and in a position to manage the client's own decision processes. All consultants are engaged on similar terms, and earn a fixed fee. One of the consultants is nominated as design leader to coordinate the whole of the design. The BPF system is not in itself a form of contract, but it does give guidance on the forms of contract which are thought to be suitable. Quite apart from its formal use in property development projects the BPF system is a logical way to approach projects of many types, and it is well worth studying. It is perhaps too early to judge the success of this system, partly due to the fact that its introduction came at a time when property development work declined dramatically in volume.

Other types of work and standard forms of contract

Mechanical and electrical engineering projects have in the past often been handled by the *Model A Form of Contract* issued jointly by the Institution of Mechanical Engineers and the Institution of Electrical Engineers. For process plant there is a *Model and Conditions of Contract* published by the Institution of Chemical Engineers. For major international projects a commonly used contract form has been that referred to as FIDIC, which is the Federation Internationale des Ingenieurs Conseils.

Special forms of contract

Many large and specialised client firms have developed their own contract forms, usually based on one or more of the standard forms published. These may however include many special requirements, as for example in the case of oil companies involved in off-shore exploration and production. In a much simpler way any client may choose to add a few special clauses to any of the standard forms in use.

Partnering

This is a means of procurement which has become more widespread in recent years, and has been developed to overcome some of the inefficiencies of traditional adversarial contracts. It has been adopted by some of the national chain stores who have set up long-term arrangements with individual contractors to carry out rolling development programmes. The concept here is that a client who has frequent need to commission work establishes a continuing relationship with a contractor to provide construction services. It has to be based on long-term trust that can be to the advantage of both parties; clients are able to predict cost levels at an early stage and benefit from a contract team which gains experience of the work expected. At the same time the contractor can more certainly plan future work-load, not subject to the vagaries of competitive tendering. Individuals begin to know and trust each other, work can be started quickly without fully detailed documentation, and costs are based on a standing schedule of rates which can be periodically updated. The method allows fast-track operation where design and construction overlap, but this does mean that good project management practice has to be followed. Apart from its origins with long-term relationships, partnering is now seen as an appropriate contract form even if it is set up for one specific project. Its dependence on trust is vital, and does perhaps imply some risk, but if it does away with some of the complex and expensive arguments about contracts then it can offer considerable benefit to all parties involved in projects. Careful initial selection of partners is of prime importance, and the detail of the contract form used should reflect the risks involved and ensure benefits for all. Partnering is still at an early stage of adoption in UK, but there is now experience both in the UK and the USA which is discussed in recent books (Baden Hellard, 1995; Ronco and Ronco, 1996; Stevenson, 1996).

Construction consortia

It is sometimes the case in very large projects that a client will place a contract not with one contractor but with a consortium of several. There are several reasons for this, including direct access to a wide range of skills and resources, but perhaps the main advantages are commercial. It may be thought that no one contractor is big enough to carry the risks of a very large project, and hence it is not possible to get a realistic bid from any one company. At the same time the client has the assurance of a large group committed to completing the work. From the contractors' point of view it gives them the opportunity to take part in large projects, while at the same time reducing the competition slightly. This approach is widely used in the case of international contracts, where it is common for at least one member of the consortium to be locally based (Stallworthy and Kharbanda, 1985; Langford and Rowland, 1995; Bennett, 1991).

PROCUREMENT OF OTHER SERVICES

All of the foregoing relates to the contractual relationships for the provision of goods and/or the construction of works. A very important part of any project however concerns the provision of services, and this also has to be formalised. Within the construction industry many services may be needed on a project, and it depends upon each project how these are procured. For example, where a project is essentially the construction of a building, then it is likely that the design will be put in the hands of an architect, who may in turn engage specialist consultants to handle the design of the foundations, the structural frame, the electrical and mechanical services, lifts and other special installations. In recent years there has been the emergence of multi-disciplinary practices which can cover all or most of these special skills. Such practices may also include quantity surveyors who would otherwise be separately appointed. The architect may be appointed as the lead consultant on a project and is then responsible for ensuring that all other design work is correctly executed. If separate design commissions are placed by the client, then these have to be properly coordinated by the client or project manager. Most of the individual professions will have their own standard terms of engagement for their work, prepared by the relevant professional institution. Recent years have seen the emergence of the 'chartered builder', someone with knowledge and experience of the building process as such, rather than the design, and such people are playing an important role in the management of construction projects.

In civil engineering and other construction work it is likely that there will be only limited work for an architect, and in such cases the principal consultant may be an engineering practice, chemical, civil, electrical, or mechanical as is relevant to the project. Where some heavy construction projects include buildings it is likely that the consulting engineer will engage the services of an architect to carry out this part of the work.

Standard terms of appointment of professional consultants

As stated above each of the professional bodies has its own standard form of agreement, or terms of appointment, and direct reference should be made to these bodies to obtain current documents. The Royal Institute of British Architects has for a considerable period used its *Plan of Work* and there are forms used by the Association of Consulting Engineers. A useful reference work for practice in USA is *Construction Management Form Book* (Cushman et al., 1983); this gives many examples, including some for the appointment of a project manager. There is also a new *Standard Form of Project Management Agreement* (APM, 1993). It is specifically for use on projects in England and Wales, and where English law applies; it is therefore not strictly applicable to Scotland, but many of the concepts included are relevant.

The terms of appointment of consultants on each project have to be clearly set out in the document appointing them. The list of possible tasks is a long one, and any one commission will include some or all of the following.

- Initial outline designs.
- Advising the client on which form of contract to use.
- Budget cost.
- Project appraisal.
- Applications for planning and other permissions.
- Preparation of bills of quantities.
- Establishing a list of selected tenderers.
- Issuing of tender documents.
- Receiving, checking and advising on the award of contracts.
- Agreeing a programme of work with the appointed contractor.
- Quality inspection of work on site.
- Authorising additional work and variations to work.
- Monthly measurement of work, and certifying it for payment.
- Authorising and ensuring payments to sub-contractors.
- Generally monitoring the terms of the contract(s).
- Arbitration (internal or external) in disputes.
- Overall management of the design.
- Overall coordination of the construction.

The project manager

Note that this list does not include project management; this is deliberate in order to highlight the special position and responsibility of this individual. It has already been stated in Chapter 10 that the project manager can be placed in a number of different situations, and may be

(*a*) a member of the client organisation itself
(*b*) the leader of the design consulting team
(*c*) the senior representative of the main contractor
(*d*) an independent appointment of an individual or a specialist project management practice.

The way in which project management will work depends upon which of these is used, and it is useful to consider again the authority and responsibility that can be attached to the role; it can be summarised in the following way. Project *management* implies exerting *control*. To *control* anything means being able to make *change*. To *change* anything implies having the *authority* to do so. Having *authority* means carrying *responsibility*. *Responsibility* involves carrying *risk*. The main conclusion to be drawn from this is that real project management can only be exerted by someone or some body that is in a position to carry risk. In most construction projects the only parties that can really do this are the client and the main contractor; any of the other parties cannot really be put in a position of risk unless they have that risk limited somehow. It is therefore a matter of major importance how the appointment of a project manager is handled and what the limits are on the liability carried. This may prompt some clients to ask the question of why a project manager should be appointed at all, especially if this increases the list of experts who have to be appointed. The answer is really that whether a project manager is appointed or not, the project will still have to be managed by someone, and if this person is not clearly identified and given the necessary powers then the project will almost certainly be less efficiently managed. It is sometimes said that the cost of good project management is between 1–2% of the total project cost, but the potential savings are much greater than this.

Where an independent project manager is appointed this may be done in accordance with the standard form of appointment issued by the APM. It is of interest to note that the standard form provides for a number of figures to be specified.

(*a*) The project manager's limit of spending authority.
(*b*) Level of professional indemnity insurance.
(*c*) Limit on liability (this may not always apply and reflects the difficulties outlined above in connection with responsibility and the client).

The Construction (Design and Management) Regulations

The Construction (Design and Management) Regulations 1994 (CDM Regulations) came into effect in the UK in March 1995 (Joyce, 1995). They require that on every project an appropriate system should be in place for the management of health and safety from the initial concept right through to final completion. Responsibilities are placed on all parties including client, consultant designers and contractors. For every project the Regulations require the appointment of a planning supervisor who has to ensure that an adequate health and safety plan is prepared and implemented. At the time of completion of the project a health and safety file has to be prepared and given to the client so that any subsequent work on the building can be undertaken with full information available. It is now clear that the planning supervisors require special training and experience. In 1995 the Association of Planning Supervisors was established through an initiative of the Royal Incorporation of Architects in Scotland and other professional bodies; it is now active in providing support to planning supervisors as they establish themselves in this new specialist role. There is clearly a strong affinity of interest between planning supervisors and project managers, and on many projects these two roles are carried out by the same office. One of the strong links is that these two functions involve all stages of a project, a factor which does not always apply to designers and contractors.

CONTRACTUAL RELATIONSHIPS — CASE STUDY

As an illustration of a procurement system it is possible to look again at the PP Ltd case, this time from the point of view of the contractual relationships for the project. Reference back to Chapter 10 shows that PP Ltd have decided to take on some of the tasks themselves, using their own staff, but engaging outside firms for other activities. Tasks undertaken by PP Ltd are

- project management for the whole design and procurement
- design and implementation of administrative and information systems
- equipment selection and plant layout
- materials sourcing
- production staff and labour recruitment
- commissioning of new plant.

Tasks where consultants are commissioned are

- market research
- product design
- organising patents
- marketing and distribution systems
- site search and selection
- building design
- product packaging design.

Contracts are placed for the supply and installation of new equipment, supplies of materials and components, construction of buildings and associated construction works, agreements with distributers, advertising and product launch. Note that the contracts listed are for use in different industries, and are likely to be different from each other in their nature and structure. The above allocation is by no means stated as a preferred arrangement, but indicates one which is possible. It is not necessarily comprehensive, but does indicate the range of different ways of working on separate parts of a project, and does demonstrate the need for coordination within one project management system, with appropriate responsibility and authority clearly defined.

Learnir

12 Project appraisal

Although this topic is being covered in one of the later chapters of this book, one of the early steps to take in any project is to make an appraisal of it in order to determine whether or not the project should proceed to the next stage. This is fairly obvious in the case of projects which are expected to make a commercial return, but it is also important in public and social projects where profit is not an objective. For any one project in the commercial field there will have to be a 'go/no go' decision before too much money has been expended simply to be wasted on abortive work. Another purpose of feasibility studies is to select from a list of possible projects the ones that offer to make the best use of limited resources. Turner proposes in *The Handbook of Project-Based Management Systems* that a 'project definition report' should result from a feasibility study which, when positive, should lead into further design work and full appraisal.

In the early days of project management it was usual to undertake a review of the financial viability of a proposed project under the heading of 'investment appraisal'. While the economics of a project are still paramount it is now general practice to carry out a project appraisal which, while including financial viability, looks at a number of aspects, including

(a) financial viability
(b) confidence in project performance
(c) competitive strength vis-a-vis other projects
(d) 'cost/benefit' analysis
(e) environmental impact analysis
(f) social and economic impact analysis

It is usually the project client who will be responsible for the project appraisal, either undertaking this task with in-house staff or using consultants specifically appointed for this task. The professional project manager is often the best person to undertake the work, which is another reason for an early appointment to this role (Hertz and Thomas, 1984).

In virtually every project there will be significant financial outlays which are expended in the expectation of later income which will exceed those outlays. The type of project where this financial benefit is not sought is where the benefits are not in terms of money, but relate to such matters as safety, comfort or convenience. It is for such cases that cost/benefit appraisal was developed (Frankel, 1990; Loasby, 1976) but it may still be valid to make a financial analysis; for example if there are several alternative ways of achieving the objectives of a project, then it is helpful to know which of them is financially most attractive when selecting one of the alternatives.

FINANCIAL VIABILITY

Some projects are amenable to a fairly simple financial assessment, for example in the case of a house-builder considering a speculative housing development. There will be several fairly well-defined costs; site purchase, design, construction, marketing, and interest on work in progress. The anticipated revenues are also fairly clear, as selling prices will probably be determined in advance, and sales will be achieved progressively during the construction phase. It is therefore a fairly simple matter to calculate an expected profit, with some small range of variation to cover deviations from anticipated costs and revenues. A point of significance here is that the time scale of the project is not very long, perhaps less than two years, and most of the relevant figures can be calculated with reasonable confidence.

Another type of project which does not require sophisticated investment appraisal is one in which there is relatively little capital invested, for example in the case of a service organisation where the only equipment is portable and non-specialised and can be readily sold or diverted to other use. In these projects there will be on-going materials and labour costs which are directly offset by revenue, plus of course some start-up costs which are relatively small. An example of an organisation of this type is a design office where the amount of capital put at risk in a new venture is relatively small.

The case of investment in industrial plant is somewhat in contrast with both of the above however, as in the case of Photo Products Ltd previously described. Perhaps the major difference is the time scale where it is likely that the plant will operate for twenty years or more, with some modification during that period. There is also the uncertainty about the market for the product both in terms of its volume and its duration. Assumptions will have to be made about these, and of course this is one of the main objectives of the initial market research, but these assumptions can be inaccurate. It has been in capital-intensive industries that much thought has been given to ways of assessing the financial viability of projects (Frankel, 1990). The chemical and oil industries have to invest very large sums of money in highly complex plants which are totally dedicated to individual products which may have limited market lives. These industries have led the way in developing appropriate techniques. A chemical company which builds a plant with a short and/or uncertain life will seek to not only recover the cost of construction but also a reasonable profit. The company will also have to think about the cost of removal of the plant once it has outlived its usefulness (Gilbreath, 1986).

There are other forms of capital investment in the construction industry which are not so vulnerable, such as the development of office blocks by property developers.

Here the significant difference is that the building is not specific to one purpose and also has a fairly long life. This means that it is likely that as an asset it will always have a value on the open market; at one time it was the case that this value could be relied upon to increase over time, but this is no longer universally true. In the times of rapid expansion of the property development industry most projects could be expected to offer a capital profit at any time; more recently it has been more important to secure an acceptable return on the investment by means of rentals.

Accounting practices in projects

Before discussing the details of the various methods of assessing project financial viability it is useful to take a brief look at the part played by accounting in project work; this falls into three distinct but complementary parts. First there is the somewhat mechanistic bookkeeping role which is not really part of the appraisal process but is worth considering briefly here. All the movements of money in return for materials, labour and services have to be properly recorded and managed. It is important to have in place good checking procedures to ensure that invoices and payments match up exactly with the goods and services delivered. Prevention of 'leakage' or 'shrinkage' are very important parts of the day-to-day control on projects, something which is made more difficult by the one-off nature of project work which changes rapidly, and which has continuous movement of people and vehicles on and off sites. It is at the same time important to ensure that suppliers and contractors are paid in accordance with the terms of their contracts and that they do not have to suffer delays or discrepancies. Another formal part of the bookkeeping role is to monitor the liquidity of the company and to ensure that at all times this is satisfactory, which can be somewhat over-simplified by stating that the sum of assets must exceed the sum of liabilities. This is a form of control which will usually be operated by the banks or any funding organisation which is lending money to the parties to the project.

The second major function of project accounting is concerned with the provision of data for a project management information system that is the basis of control of the project. It has already been pointed out that money is often used as the unit of measurement of work on a project simply because it is common to all inputs (since they all have costs), and because the data is probably being kept in great detail for the bookkeeping purposes mentioned above. It was also stated earlier that this use of money as a measure has its shortcomings, but it is nonetheless widely used. There is of course a great deal of useful cost information relating to the early stages of a project which can be used to modify the planning of later stages in order to satisfactorily complete the project. It can also be used in the planning of other similar projects. This type of accounting is sometimes referred to as 'internal' accounting, since it only affects the internal workings of the project organisation. It contrasts with the bookkeeping or 'external' accounting in a number of ways; for example in external accounting accuracy is essential even if it takes time, whereas in internal accounting speed is usually more important than accuracy. It would clearly not be acceptable to base the sums payable to operatives or suppliers on approximate measures, but it is good enough to do so when planning the next phase of the project, which may in any case have a number of differences in detail.

The third aspect of accounting is in relation to the financial appraisal of projects and there are a number of different ways of looking at this:

Return on investment

This method, which is sometimes abbreviated to ROI, expresses the expected annual profit as a percentage of the total sum invested in the project. While this is apparently a very simple measure it is complicated by a number of factors.

(a) It assumes that the profit yield is the same in each year and can be represented by a single figure. In reality there may well be both predictable and non-predictable variations in the level of profit. Furthermore the quoting of a single annual figure gives no indication of how long the yield will be produced; a project with a 3-year life will look the same as one with a 20-year life, if the annual rate of return is the same.

(b) A clear statement must be made as to whether the profit figure quoted is before or after tax, and any assumptions about work in progress and materials in stock must be clarified.

(c) Normal profit calculations make an allowance for depreciation. It must be made clear whether this has been done and whether the rate is that allowed by tax authorities, or some other rate which may take account of the risk of the project.

(d) Government and similar grants towards capital investment also complicate the calculation of the actual capital invested.

(e) ROI takes no account of the time value of money, and will give equal weighting to sums received at the time when the project facility comes into operation, and to sums received in say 10 years' time. It is possible to overcome this by deducting from the profit percentage a figure corresponding to the interest rate that could be earned if the capital sum were placed in a broadly similar investment.

(f) Perhaps the most important drawback to ROI is that it does not indicate the magnitude of the investment. If the investment is being made to achieve a particular purpose then it might well be preferable to invest £100 000 with an expected ROI of 8% rather than £500 000 with an ROI of 10%. On a direct comparison of percentage return then the latter would appear to be best, but if consideration is given to the magnitude of the sum at risk then the former and much smaller investment seems to be the better way to achieve the given objective.

All of these factors will make an overall decision somewhat uncertain, but if a range of similar projects is being considered and the same assumptions are made about them all, then comparison between the alternatives is reasonably reliable. The advantage of ROI is that the calculation is fairly straightforward and comparison of the alternative projects could not be simpler.

Pay-back period

This is an extension of the return-on-investment method, and consists of dividing the expected annual profit into the capital cost of the project in order to obtain the number of years which will be required to 'pay back' the capital sum invested in the project. This does partly overcome the objection to ROI of taking no account of the magnitude of the capital sum, but it still does not state what that sum is. While it does give the number of years before the investment is recovered it does not indicate how many years thereafter the project is likely to run and hence what 'surplus' recovery of capital might be achieved. As in the case of ROI it takes no account of the time value of money, nor does it give any indication of the risk of the project.

Many of the shortcomings of the two methods indicated so far can be overcome by looking at cash flow methods, and these can be further refined if the cash sums are discounted in order to take account of the time value of money, namely the fact that a sum of £100 received now is very roughly worth the same as £110 received in a year from now. This has led to the technique known as 'discounted cash flow' which is inevitably more complex to calculate than either ROI or pay-back period, but does give a much more meaningful representation of the value of the project. Associated with this is the concept of 'net present value', and these two approaches are now considered together.

Discounted cash flow and net present value

These two methods are based on the two concepts of recording all cash flows in and out of the project and then discounting their value to take account of the time value of money. They ignore depreciation since this is not represented by a cash flow, but they can take account of tax, maintenance and repair because these are paid for in terms of a cash out-flow. The procedure is first to set out all the anticipated flows of cash relating to a project and the dates on which they will take place. In order to do this it is necessary to look some way into the future. In fact, in the case of setting up a production facility, it is not only the construction and equipping of the plant which must be considered, although this is what we would normally consider to be the project, it is also necessary to anticipate the whole life of the plant, together with any upgrading or refurbishment which might be undertaken. This is the approach referred to in production management as life-cycle costing, where all the costs associated with a plant and its products are taken into account over the whole of its working life. It is important to note that a very similar phrase is used in project management to denote a rather different thing; this is the recognition of a 'project life cycle' which goes in turn through the steps of project concept, project scope definition, project appraisal, design, tendering, construction and/or installation, commissioning, hand-over and close-out. Note that in this context the continuing operation of the completed facility is not included and the project life is restricted to the 'concept to completion'. This is in some ways analogous to the idea of a 'product life cycle' which usually covers the period of invention, design, production planning, launch and manufacture up to the point where production is closed down or the product is superseded by an updated or new model. All this may seem a little confusing and therefore a brief summary may be useful.

- 'Project life cycle' covers from concept through to completion of construction and/or equipping, and possibly including the launch of a new product.
- 'Product life cycle' is the period from invention through to cessation of manufacture and selling, but not the end of its use.
- 'Life-cycle costing' embraces all the costs of owning and operating a plant, a building or a piece of equipment up to and including the scrapping and demolition stage.

Project appraisal using discounted cash flow

Here we are concerned with the last of the three definitions above, since we are evaluating the viability of a new facility over its working life. The way to approach this is to set out a list of all the cash flows, in and out, which are directly attributable to the plant, constructed facility or equipment being assessed. The timing of all these anticipated flows must also be noted. It is recognised that while it is possible to be fairly accurate with the initial capital costs, the estimation of cash movements in years to come will be less accurate; it will be seen however from the following calculations that cash flow sums several years into the future have a relatively small impact on the overall figures, and accuracy is therefore not essential. It is of course true that any method of financial appraisal will depend on approximate estimates of future cash flows, and discounted cash flow (DCF) is perhaps the least vulnerable to inaccuracy in these estimates.

The concept of discounting future cash flows

The concept is that the timing of payments affects their real value, and that £1000 received today is worth more than the same sum received in a year's time; the £1000 if available now could be invested and earn interest of say £60 within a year, which would not be available if the £1000 were received in a year's time. Look at the same concept from the viewpoint of a company which is due to make a payment of £1000 12 months from now. If the company placed a sum of £x in an account which pays 6% p.a. it could have £1000 at the end of the year to meet the payment, and it is possible to work out the value of x from the equation

$$x \times \frac{106}{100} = 1000$$

hence

$$x = \frac{1000 \times 100}{106} = 943.4$$

It can therefore be seen that £943·40 invested at the beginning of the year at 6% interest would have a value of £1000 by the end of the year. It is then said that the £1000 due in a year's time has been discounted at a rate of 6%, and has a 'present value' of £943·40. The same calculation can be carried out for any rate of interest, and for any number of years. Clearly this becomes a tedious calculation, but fortunately all the work has been done in creating Table 24 of 'discount factors', which is shown in the appendices. In this it can be seen that the discount factor for

6% for 1 year is 0·943, i.e. the value calculated above. If we had another payment due in 5 years and wished to apply a discount rate of 10% then the factor from Table 24 is 0·621. It can be seen from these figures that the present value of sums due a few years into the future drops off rapidly with increasing years and increasing interest rates. This can be illustrated by a very simple example of someone being offered two alternative methods of payment for completed work

A: £500 paid immediately

B: £100 paid at the end of each of the next 6 years.

Using the table of discount factors in Appendix 2 the two can be compared in Table 9 on the basis of a discount rate of 10%:

The present value of payment method B is only £435·40 despite the fact that the sum total of cash paid is £600. The present value of payment method A (immediate) is the full £500 and is therefore to be preferred by the recipient. If the discount rate applied was only 6% then the present value of the 6 years payments would be £521·7 which would then make method B slightly better for the recipient.

Selection of an appropriate discount rate This is a subject of great debate, and is one which is usually solved by judgement rather than by calculation. It is unlikely that a rate lower than the current rate of interest for loans would be applied, since the concept of discounting really derived from the notion that it would be better to invest money than go to all the trouble of creating a large project to earn a lower rate. There is a strong case to be argued that the discount rate should always be significantly higher than interest rates to compensate for the risks involved in a project. In the 1980s it was not uncommon for discount rates in the range 20–30% to be used in the case of high-risk projects, perhaps where it was not possible to see a market for the product very far into the future. The actual value selected may have a large influence over whether or not development is justified at all, but if several broadly similar projects are being compared with a view to selecting which one will proceed, then the value is not too important as long as the same one is used for all the candidate projects.

Table 9. Comparison of cash flows using a discount rate of 10%

Year	Sum payable	Present value
1	£100	0·909 × £100 = £90·9
2	£100	0·826 × £100 = £82·6
3	£100	0·751 × £100 = £75·1
4	£100	0·683 × £100 = £68·3
5	£100	0·621 × £100 = £62·1
6	£100	0·564 × £100 = £56·4

Total present value = £435·4

Discounted cash flow for the Photo Products Ltd case study

The categories of cash flows listed must include all those which are expected to occur in the life of the plant, and listed below are those which might apply to the general case of a project which comprises the construction, equipping and operation of a new factory or plant for the production of manufactured goods. The list is therefore set out as it might apply in the case of Photo Products Ltd, the case study that has been used several times earlier in this book. The convention used here is that a cash flow 'out' is a flow laid out by the client and invested in the project. A cash flow 'in' is a flow of cash into the client's central account from the project.

Cash flow out The purchase of the site — this will usually be a single payment out, but it may be complicated if money is borrowed to fund this purchase, in which case the repayments may be phased or deferred until some later date. Care has to be taken to ensure that interest on money borrowed is treated in a consistent way throughout this analysis, and is really the subject of specialist attention by an accountant. Perhaps the simplest approach to handling interest is to regard it as a cash flow out at the time of payment; however if the initial cost is funded out of the client's own resources no direct interest is payable. In these circumstances it is necessary to make a notional interest charge to the plant so that a fair comparison can be made with the case of borrowed money. The same consideration should be given to any money borrowed for the purchase of capital goods for the plant, as listed below.

The design and construction of the building — most forms of contract would stipulate phased payments during the construction, based on actual progress, and there is often a final payment to be made once it has been agreed that every item has been completed and properly measured. Often a small percentage of the contract sum is held in retention for a stated period, often 1 year, to act as a form of bond to ensure that any defects which come to light are corrected.

Note that in some situations the factory might not be built or bought by the client company but rented from some other organisation. In this case there would be no initial outflow of capital for the purchase, but a regular quarterly or annual payment, which might increase with time on some calculated basis, e.g. inflation-linked. There could however be initial costs of taking over and making modifications.

The design and installation of the manufacturing equipment — this is likely to be an accumulation of many separate items bought from a range of suppliers, but probably all purchased at about the same time, when the construction of the building is at or near completion.

A wide range of administrative and organisational activities will have to be completed at an early stage, some of them in fact before the main project go-ahead is given. These will include market research, product outline design, patent searches, production design, and others. All of these will cost money in terms of the working time of employees and consultants, and will arise in the early stages of the project. Similar tasks will have to be completed later, but before production and selling can begin, such as advertising material, staff recruitment, materials and components purchasing contracts, distribution and servicing networks, information systems for controlling manufacturing and selling, and a range of jobs associated with establishing a new production facility and its support. The timing of these costs will be over the period of construction of the building and the installation of the plant, and also during the commissioning and production build-up phase.

Note that the costs so far are really all concerned with the project of getting the plant ready for operation; the next stage is the commencement of steady-state production. While it would be feasible to enter all the cash flows, both in and out, which are associated with production, namely materials and components, labour, energy and other services, transport, production planning and control, selling and distribution and so on, this would be somewhat tedious at this stage. It would be usual to estimate an operating account for the plant and to simply take the figure for operating profit or loss as a cash flow in or out respectively. Dealing with tax is complex at this point, and it is probably better to regard this as a charge on the client company as a whole, rather than on the specific plant, and therefore not to enter it as a cash flow out.

While routine maintenance should be regarded as a production operating cost, there will from time to time be major items of refurbishment or updating of plant, or even its total replacement to make a different product. The building may also have to be upgraded at some stage. This could represent a very considerable further investment in the plant, but it will arise some number of years into the plant's working life, and will consequentially be discounted.

Cash flow in   Depending upon the time and the location there may be various forms of development grant available. These will usually be either phased or paid once the plant is in production. The availability of such grants has in the past been a major factor in the locating of new projects, and in some ways has distorted the distribution of industry. Much of this has now been sorted out, but it does still represent a question mark in some developments.

The major in-flow of cash to the company deriving from the project plant is in the form of operating profit, and this has to be carefully estimated. It is quite possible that this will not be a uniform flow, because of the absence or presence of competition, but a forecast has to be made.

Cash flow in or out   At the end of the working life of the plant it will have to be shut down and scrapped, and the net cost of this could be positive or negative, depending upon a number of factors. In the PP Ltd case it is probable that there would be a net cost of scrapping the equipment and removing it from the site. The building may have some residual value for the manufacture of another product, or for another company altogether. It is likely that the site itself will have a residual value and will not be badly polluted. These cash flows in or out are very uncertain when looking 20 or more years into the future, but it must be remembered that precisely because they are far into the future they will be heavily discounted.

Special cases of plant removal can cause huge problems in some industries, for example, in the case of a nuclear power station or an off-shore oil platform. In cases such as these the shut-down and removal costs are extremely high, as has now been realised in these two specific industries. Questions have been asked as to why no thought was given to building into these installations the means of making them easier and cheaper to remove at the end of their working lives. There is one answer to this question namely that thought was given to the problem but it was decided that no action should be taken. This was in part due to the fact that an optimistic view was taken of the ultimate removal costs, but more importantly the discounting of removal costs over a long period at a time when interest and discount rates were high meant that any savings that could be made in the removal costs had hardly any value when discounted to present value. For example, at a discount rate of 16% over 20 years the discount factor is 0·051, namely down to about one twentieth of its cash figure. To achieve this saving, initial costs would have been incurred at the construction stage, i.e. at their full cash value, and it was difficult to justify such expenditure when the return in 20 years' time looked to be so low. This reasoning does cast some doubt about the relevance of DCF methods on projects of this type. While the purely financial argument may still be valid it does have the effect of influencing the initial design in such a way that final removal is not provided for. The problem still remains however, and the actual 'cost' in work-hours of people and equipment will be high at the time of removal. It would perhaps be preferable in such examples to modify the approach, and use a much lower discount rate which would roughly equate the net rate that could be earned from a safe investment.

The special case discussed above does not by any means invalidate the DCF approach to most projects, but simply acts as a warning that it must be used with careful thought. In many industries the problem of removal is not a great one, e.g. it is cheap to actually discard computer equipment or to throw out a series of documents, provided of course that provision for their replacement has been made. In most cases DCF is used to compare a series of alternative projects, and in such cases even if the discount rate chosen is open to question it does not matter greatly if the same figure is used in all of the alternatives studied. Comparison between alternatives of a different financial structure would have to be more carefully considered, for example in choosing between the capital outlay on a new site and building compared with a long-term rental agreement for a physically similar facility.

Solution rate of return and net present value

This is a specific single-figure measure of the financial viability of a project. It consists of writing down all the cash flows in a project and then discounting them in terms of an unspecified discount rate, say s, taking account of the timing of these separate flows. These are then summed into the form of an expression in terms of the unknown s, this expression is then equated to zero and solved for s. This gives a value of the discount rate which would have to be applied in order that the project should exactly break even over its whole life. It is then possible to compare alternative projects and/or straightforward cash investment simply on the basis of one single factor, the solution rate of return. As a first approximation measure this method has benefits, but it is thought by many to be too simplistic; it does mean however that it is fairly easy to identify those projects in a list of alternatives which are worthy of further study. One means of such further study is to examine the net present value (NPV) of a project at all stages throughout its life as set out below.

Here the procedure is to set out all the cash flows as before and to discount each of them in accordance with the selected discount factor and the timing of the cash flow. The next step is to plot these in the form of cumulative value to a base of time, as shown in Fig. 57. There are two main advantages of this form of presentation. It is easy see just how far each project goes 'into the red', i.e. what is the measure of risk if at any stage the project life is cut short by some unexpected event. Also, in the comparison of alternatives it is possible to see how they would perform relative to each other throughout their expected lives, not only at the end, e.g. which one is the first to have a positive NPV. These are illustrated in the simplified example shown in Table 10 and Fig. 58, in which the following assumptions are made.

(a) All cash flows in and out take place at the end of the year.
(b) The discount rate is 10% p.a.
(c) The initial capital outlay in both projects A and B is spread over 3 years (note 1 in Table 10).
(d) Project B has a more complex plant than A at 48% higher cost.
(e) Project B has more efficient plant and provides 100% higher profits, but deferred 1 year (note 2 in Table 10).
(f) Both projects have the same ultimate scrap value of £2300, being the site value (note 3 in Table 10).
(g) Discount factors have been rounded to two figures for simplicity.

The detailed calculation of cumulative net present value of the two projects A and B is set out in Table 10.

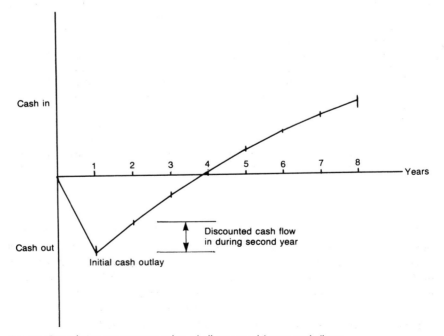

Fig. 57. Cumulative net present value of all expected future cash flows

Table 10. Cumulative net present value of projects A and B

Project A: discounted at 10% p.a.

Year	Cash flow			Discount factors	Net present value		Notes
	Out	In	Net		Discount value	Cumulative value	
1	500		−500	·91	−455	−455	
2	800		−800	·83	−664	−1119	
3	1000	100	−900	·75	−675	−1795	1
4		300	+300	·68	+204	−1590	2
5		300	+300	·62	+186	−1404	
6		300	+300	·56	+168	−1263	
7		300	+300	·51	+153	−1110	
8		300	+300	·47	+141	−969	
9		300	+300	·42	+124	−843	
10		2600	+2600	·39	+1014	+171	3

Project B: discounted at 10% p.a.

Year	Cash flow			Discount factors	Net present value		Notes
	Out	In	Net		Discount value	Cumulative value	
1	1000		−1000	·91	−910	−910	
2	1200		−1200	·83	−996	−1906	
3	1200		−1200	·75	−900	−2806	1
4		600	+600	·68	+408	−2398	2
5		600	+600	·56	+336	−1670	
7		600	+600	·51	+306	−1364	
8		600	+600	·47	+282	−1082	
9		600	+600	·42	+252	−830	
10		2900	+2900	·39	+1131	+301	3

It can be seen from both Table 10 and Fig. 58 that project B has a higher final value than does project A at today's net present value. However it has a much higher 'maximum risk' value at the end of year 3, and is in fact at greater risk through 9 of the 10 years of the life of the plant. It would perhaps be difficult to persuade a client that it was worth investing the larger sum of money in order to increase the profit level. It is worth noting that the solution rate of return would be almost exactly 12% in both cases, and that the use of this simplistic statistic might give little or no guidance in this example.

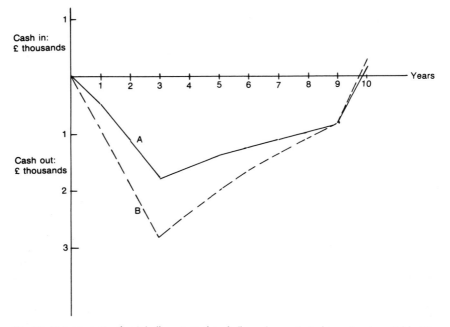

Fig. 58. Net present value of all projected cash flows for project data set out in Table 10

Selecting an appropriate discount rate   There has always been a great deal of debate about this question, which is very much one which has to be tackled by the people who are considering the investment. If the project is to establish a plant to manufacture a product it will be a matter of trying to see how far ahead the market for that product can be relied upon. If the market can be seen to exist for many years then it is reasonable to select a rate which perhaps reflects the cost of borrowing, but if the future can only be seen for very few years, say five, then it will be necessary to effectively write off later years by using a high discount factor. In some risky projects discount rates as high as 30% have been used. Some consideration must be given to inflation; this can partly be done by making some assumptions as to how this will affect future trading and hence will be reflected in the cash flows, and partly by making an allowance in the discount factor. The whole issue can become somewhat complex and is really outside the scope of this book.

Sensitivity analysis   It can be seen from the example detailed in Table 10 and the comments upon it that the interpretation of the calculations can be affected by the assumption of a value for the discount rate, i.e. the outcome is 'sensitive' to the value of discount rate chosen. There are many other factors which can influence the appraisal of a project, including

(a) the market assumed for the product of the plant, both in terms of volume and price
(b) the expected life of the manufacturing plant

(c) the operating cost of the plant and the prices of materials and labour

(d) the cost of construction of the buildings and installation of equipment

(e) the completion date of the project and start of production, since even a small delay may cause the loss of the first year's market, and this may never be recovered.

For each of these factors it is possible to carry out a recalculation of the NPV of a project for a series of assumed changes in the factor, and then to construct a graph of the form shown in Fig. 59. This shows the effect on NPV of percentage changes in the factors listed above. In the case shown, increases in either the plant life or the market have a small positive impact on NPV, which is therefore not very sensitive to these two factors. Operating cost increases will have a small negative effect, but the construction cost and completion date are both shown to have a big negative impact, and so it can be concluded that the viability of the project is very sensitive to these two construction-related factors. In a project such as this one it would be expected that great pressure would be exerted to bring the project to completion on time and within budget; clearly there is a need for good project management control. In any project it can be determined which of the factors are most sensitive, and an attempt can be made to improve the accuracy of the forecast. This will usually mean spending further effort and time in making the estimates, possibly carrying out analysis or even physical tests, and more detailed market studies simply to improve the quality of the information on which the appraisal is based.

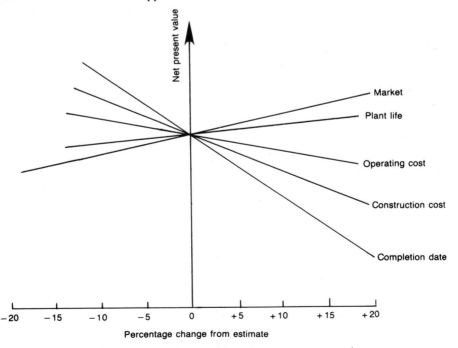

Fig. 59. Sensitivity of net present value to changes in input parameter values

COMPARISON OF ALTERNATIVE PROJECTS AND SELECTING WINNERS

Commercial projects are usually compared on a financial basis taking account of the availability of money and the alternative uses to which it can be put, even if it is only simple investment. Often there will be many projects within the development department of a manufacturing company all seeking funds in order to go ahead to the production stage. There may not be funds or other resources available for more than a few of the competing candidate projects, and a selection has to be made. One of the ways of dealing with this situation is to list all current projects in descending order of financial benefit, perhaps in terms of NPV or whatever terms seem appropriate. Funds are then allocated down this list until there is no money left. An alternative to this, quoted by a manager in the chemical industry, was to start by striking off the list the project at the bottom, and then working up through the list deleting everything 'until the screams get too loud'. This may not seem elegant, but may be quite effective, and it does also identify those projects where there is a strong project champion or product champion.

ENVIRONMENTAL IMPACT ANALYSIS

Environmental impact analysis (EIA) has become standard practice in the case of major construction such as motorways, quarries and mines, power stations, oil refineries and similar projects which have a potential impact on the environment, to carry out a full study of this likely impact. This may be simply a requirement that such a study is carried out by a competent professional, or in the case of large projects it may be the subject of a public enquiry. Not only large projects are amenable to environmental analysis, but this should not mean a full public enquiry in every project, simply that environmental issues should be considered as a part of the normal planning consent. Sometimes the phrase used is environmental impact assessment, which is also shortened to EIA (Jain and Hutchings, 1978).

In order to achieve impartiality, it is usual for an EIA to be undertaken by or at the behest of a government or local government authority. This means that it will seldom be the responsibility of the project manager to carry out the assessment, but it is very important that the project manager is aware that it is taking place, and can take account of its possible effect on the progress of the design and construction. The whole subject of EIA has become somewhat emotive and confrontational, as it is seen by some project clients as a further hindrance to development, often causing delays and additional costs. This need not be the case, as is explained below. For this reason it is useful for the project team to be fully conversant with the process of undertaking an EIA.

Purpose of environmental impact assessment

The purpose of EIA is to examine in advance of the construction of a new facility the impact it will have on the environment, with a view to reducing as far as is possible any adverse effects. In the extreme case where these adverse effects cannot be kept below acceptable levels it may be the case that the EIA will conclude that the project should not proceed at all. In most cases this should be avoidable if the EIA is undertaken while the decision-making processes are still in hand, for example in the selection of a suitable site, or in the design of the buildings and process plant that will be constructed. Many of the potential sources of pollution can be removed or reduced by appropriate design, and it is entirely logical that environmental factors should be just as much a part of the design as are ground conditions, wind-loading and so on. If the design phase and EIA are carried out in parallel there should be very little reason why environmental considerations should cause any significant delay to the project; the EIA should not be treated as an extra stage which has to be executed in sequence, but as an integral part of design. Retrospective action on the design because the EIA was not prepared in parallel with it will be expensive in time and cost. If the assessment is made post-design and is critical there may be a temptation to ignore it if it is not actually enforced by planning authorities. This could in time lead to additional expense if it is necessary to modify or even shut the plant down at a later date. Where the EIA has been carried out independently of the decision making it is often the case that the outcome is simply either for or against the project. What happens when the view is against is that the project is stopped; when the view is for, the project may proceed with a number of environmentally objectionable aspects — either way there is a loss to the community. It is now thought that it is far preferable to treat EIA as an integral part of the decision and design processes. The outcome of the assessment should be a balanced design which takes account of all requirements, those of production, investment, safety, social value and environment.

Content of an environmental impact assessment

A useful approach to assessing impact is to regard the proposed project as a 'system' and draw an imaginary boundary around it. All the 'transactions' across that boundary can be meticulously listed and quantified. These will include both desirable and undesirable inputs to and outputs from the system; it is largely the undesirable outputs that will be the concern of the EIA. The factors which should be examined include

- effluent discharged to the local drainage system
- potential effluent and solids leaked to the ground
- gaseous and air-borne discharge to atmosphere
- noise and vibration
- radiation emissions
- risk of explosion or other catastrophe
- extraction of water and other materials from the site
- traffic movements in and out of the plant
- visual impact of the constructed project
- the potential for linked development which may be undesirable, e.g. downstream processing.

SOCIAL AND ECONOMIC IMPACT

The discussion of social and economic impact is a topic which has only recently been fully aired. The financial appraisal discussed earlier is concerned with company economics, what now remains to be looked at is the impact on the local economy in which the project is being placed. This has traditionally been the responsibility of local planning authorities, but their objectives have perhaps not been wide enough and have concentrated too much on purely local issues. The following discussion is more of an agenda for consideration than a recommended course of action. Use is often made of cost/benefit analysis to quantify factors that do not immediately lend themselves to purely financial criteria (Loasby, 1976). Any social/economic appraisal of a major building development should address a list of questions.

(a) To what extent can market forces be allowed to have a major influence on broad planning issues?
(b) Are the most appropriate criteria being used in decision making by developers?
(c) Are the most appropriate criteria being used in planning decisions?
(d) Are planning and development issues being decided by the most appropriate people?

Before seeking direct answers to these questions, a number of more specific topics are discussed.

Use of greenfield sites for industry

The post-war development of new towns had the objective of creating both good working and good living environments to make possible the regeneration of industry and the improvement of life for many people. There were both failures and successes, but that is not the issue here. One of the outcomes in more recent years during times of recession or stagnation has been that some of the new towns have not been able to easily meet their expansion targets, and as a result have entered into strong competition with other areas to attract new industrial development. This competition has not only been between the new towns themselves, but between new and old. Often the very favourable terms offered by new towns, including rent and/or rates holidays have tempted existing companies to move from old to new sites. This has resulted in both the much-publicised 'new jobs created', and also in the much less publicised jobs lost. It is true that the new jobs may be in much 'nicer' surroundings, but they have not always turned out to be in the most happy surroundings. The new towns have taken some considerable time to settle down to a social mix which includes people of all ages and groups.

Another aspect of new town development competition has become apparent in the attraction of inward investment from overseas. It is a direct result of the fierce competition between both old and new areas that the selection of a location by an in-coming company will include not only economic criteria but also factors relating to the 'desirability' of the physical environment in the area. The result of this is that regions offer attractive greenfield sites, in the knowledge that if they do not, some other area will make such an offer and be successful. This means that greenfield sites are used while derelict in-town sites remain unused, with the consequence that concrete deserts are left in towns. An example of this is the area close to major rivers, where many new industries have been established on the downstream banks on attractive green sites, while nearby there are many derelict open spaces, despite the fact that both areas may have the same ease of access to the motorway system, the railways, an airport, and to residential areas for employees and both places often fall within the same local authority.

Use of greenfield sites for retailing

There has been much recent discussion on the pros and cons of retail sites outside built-up areas. There has clearly been great demand from national retail companies for good large sites with big car-park areas, and many of these have been located on green sites close to towns with access very largely dependent upon customers with private cars. Again these developments do not provide additional retail outlets with additional jobs, but rather create a transfer of jobs from one area to another. One of the results of this is to denude the shopping areas of towns, in particular those small towns which are close to major conurbations. This denuding of town centres has many serious consequences. Local people without cars are not able to get out of town and have been deprived of in-town shops. Many smaller specialist shops depend upon being close to the big stores in order to obtain enough custom; they are then at risk of closure and are lost as a service to the public. The overall result is that retailing is more and more in the hands of the big groups, with a consequent loss of choice by customers, and all the other potential hazards of oligopolies. Questions are now being asked in many places about our growing dependence on the motor car, and there is much pressure to restrict it in some way. If there is a significant reduction in car use there will a very serious reduction in the viability of out-of-town shopping sites, but it may be very difficult to contemplate relocation in town centres.

It would appear from the above that we are making fundamental changes to our use of land without taking full account of the long-term social and economic consequences, and all in the interest of short-term benefits. Furthermore the financial structures are such that no investment reserve will have been made which can be drawn upon to fund any further changes which may be necessary. This concept of providing funds for work in the future is foreign to many developers, but is now being thought about in some of the heavy engineering industries where future (but as yet unidentified) work will have to be carried out to meet liabilities arising from the original construction.

The process of procurement of buildings

The traditional procedure for procuring a building is either to buy or rent from a developer an existing building (or one already under construction), or to commission one to be built directly. For the purposes of this chapter renting or buying are regarded as being largely the same, and suffer the same drawbacks, while the

213

commissioning of a new purpose-designed building does have both advantages and disadvantages. The advantages of obtaining a building from a developer will include

- immediate or early occupation
- access to a wide range of good locations
- the developer's expertise in the commissioning and construction of buildings
- a wide range of financial advice and support.

By comparison the advantages to a building owner of commissioning a purpose-designed building will primarily be the following.

- Involvement in the briefing and designing of the building. Where any special requirements hold this can be a big advantage, compared with taking over an existing building with the added costs of any necessary alterations.
- Depending upon particular circumstances there may be financial advantages.
- While not of direct benefit to the user of the building, there may be benefit to the community in which the building is located.

The case of purpose-designed structures is rather similar to what happens in the civil engineering industry, where in almost every case the structure is custom-built to meet the particular needs of the project, with the design carried out totally with that project in mind; the design may indeed be carried out by the client organisation. The client will be responsible for the operation, maintenance and ultimate disposal of the structure, and probably for the subsequent use of the land on which it was built. This contrasts with the case in which a developer erects a new building and concurrently seeks a client to buy or rent it. Given the long time scale to design and build it is most unlikely that the user's needs can be fully taken into account, and hence the users get a building which is less suited to their needs than would be wished. This separation of user from designer necessarily introduces a communication barrier, which can be made more obstructive if a further separation takes place, e.g. if the developer sells the building to another party such as an investment company or pension fund.

Criteria affecting the design of a building

There are many factors which will be taken into account in the briefing and design of a new building, including both the technical requirements and the economic assessment. There will always be a trade-off between what is desired and what is affordable. An important area of study however is the trade-off between initial construction costs and the costs of both maintenance and use of buildings. For a long time this trade-off was ignored, since capital costs and revenue costs came out of separate budgets which were not interchangeable. Furthermore maintenance costs and operational costs were similarly separated, coming from separate pockets. Much work has been done in this area with the introduction of life-cycle costing, where the sum of all costs incurred in the ownership, maintenance and use of a building over its entire life are taken into account. This has helped to bring about some improvements in building design, e.g. in relation to thermal insulation, but has not yet realised its full potential.

If DCF methods are used in appraisal a high discount factor leads to effectively ignoring the costs of long-term liabilities; it then follows that there can be no justification for spending money at the front end of a project with a view to saving money later in the life of a building. This applies only partly to running costs, but much more importantly to mid-term upgrading costs or ultimate removal costs. There are now problems arising in some heavy engineering projects where ultimate removal was largely ignored at the time of initial construction due to its very high initial cost.

Buildings may not be in the same position as these heavy structures, but there are similarities. What is evident in buildings is that short-term initial building costs have dominated the design of many buildings, especially those erected in the 1960s. It is worthy of note that the construction industry found a lot of its work in recent years in the refurbishing of buildings put up in the 1860s and those put up in the 1960s. Examples of this would be the Albert Dock scheme in Liverpool, compared with 1960s multi-storey housing. This is a reflection of the emphasis put on short-term factors, especially the urgent need for public housing in the post-war period.

The solution to these problems may lie partly in re-thinking how we evaluate new building proposals. In particular more thought must be given to the selection of discount factors; either by using a high value for future income and a lower value for future expenditure, or by using some other (non-inflatable) measure of value than money, perhaps a man-hour of building workers' time.

It may be concluded that the questions posed at the start of this section (see page 212) on economic and social appraisal can be answered as follows.

(*a*) Market forces can only be relied upon to indicate the total demand for new buildings, and action should be based on a longer-term planning strategy.

(*b*) Developers may currently make use of criteria which are most appropriate to them, rather than to building users and the public at large. Other criteria should be used, e.g. the long-term demand on resources. Discounted cash flow methods should be reviewed in this light.

(*c*) Planners may need to take account of future liabilities on society, brought about by their decisions. This may mean requiring developers to incorporate long-term features, by using different assessment criteria.

(*d*) A balanced view is required, and this must come from cooperative effort by entrepreneurial and regulatory groups. Treating all developments as long-term projects with appropriate project management control would yield overall benefits.

13 Variability, uncertainty and risk

The concepts covered in this chapter are perhaps the most difficult areas that have to be dealt with in project management. Unfortunately it is also a fact of life that they exist and must therefore be tackled. It is also true that they enter into the management of projects to a greater extent than in most other areas of management. It was quite significant that more than 30 years ago in 1966 the Tavistock Institute published a report on the building industry entitled 'Interdependence and Uncertainty'; the title was prompted by what the study team found — that there was a great deal of interdependent activity in the industry, something we recognise now in project management. They also found that there was a great deal of uncertainty, and furthermore concluded that the industry actually thrived on that uncertainty. It is perhaps no longer true that the industry thrives on uncertainty; ways have now been found to deal with it, and clients are expecting it to be dealt with properly.

VARIABILITY

The understanding that variability is inherent in almost everything around us is somewhat difficult for young engineers and others who are accustomed to dealing with things in a deterministic way. Engineers, especially those who are at the immediate post-education stage of their career have the notion that there is a right answer to every problem, and this can usually be expressed in terms of an exact number. This idea is soon dispelled when those concerned with production start to look at quality control processes, and examine the data which is gathered by measuring the dimensions of mass-produced items which are supposed to be identical. It will be found that a variation exists from one item to the next in an apparently random manner. The processes of quality control in fact are based on recognising the point at which the variation is no longer random and there is some underlying cause of the variation. Production managers have to learn about the variation of the process and how to use it in the control of that process. It is worth noting that data could be collected not only in terms of a linear dimension of a machined component, but also the crushing strength of bricks or samples of concrete, the weight of material in pre-packed bags, and fitting tolerances in building components such as windows and doors. The question of dealing with variability in quality management has been discussed in Chapter 7; it is mentioned here only in terms of illustrating the variability of processes.

Variability in project management

This manifests itself in a number of ways which make it more difficult to deal with than in the management of manufacturing production. First there is the fact that in any one-off project the actual work varies from day to day. Each day a project will bring the completion of some activities and the commencement of others. This is of course one of the most important differences between project management and production management, and means that forward planning is both more difficult and more important in the case of projects. Even where the same skills are involved in project work from one day to the next the actual task may be different, or perhaps in a different location, under different working conditions, or carried out by different individual operatives. Each of these factors will mean that both the standard of work and the time taken to do it will vary from one item to the next. Quality variation has already been discussed — it is also necessary to consider the variation of the time taken to complete a task, with the consequent effect on other succeeding tasks; this is dealt with in some detail later in this chapter. There is also the effect on costs and their control. It can therefore be seen that not only is there the natural variation which is inherent in repeated processes, but there is an added dimension to the variation caused by the nature and environment of the work. Planning ahead of project-based work is indeed both more important and more complex, especially in large one-off projects.

UNCERTAINTY

There is a strong link between variability and uncertainty, if only because we cannot be certain about the extent and manner in which some factor will vary. There is however another dimension to uncertainty, and that is the question of whether or not a random or chance event will occur, e.g. a flood or an accident, and that is something we have to deal with in projects. It may be helpful in the context of project management to state what is meant by certainty, since very few events are absolutely 100% certain in any context. In a well defined project such as the Photo Products Ltd case it can be stated with 'reasonable certainty' that every activity will have to be completed; drawings will be prepared, contracts awarded, site preparation completed, buildings erected, equipment installed, operating systems established, information systems installed and production commenced. Their exact form may not be known in advance but we can be certain that they will have to take place, and it is possible to think in terms of dealing with variability in these cases. Other possible events in that project will however be uncertain, such as the possibility of a fire during construction, the sudden bankruptcy of one of the contractors or suppliers, a legal case on patent rights. Any of these may or may not arise, but if they do the impact on the project could be very significant. Such events have to be treated somewhat differently — a possibility could be ignoring them and hoping for the best. Good project management would however require that these random events should be fully considered and dealt with in a logical way. This then becomes the subject of risk analysis and risk management, the third concept in the title of this chapter.

RISK

The above paragraph shows that dealing with uncertainty leads us to consideration of risk in projects, and consequently to the management of risks. Several definitions of risk are used, and mostly refer to hazard or chance of loss, but there are two aspects to risk which must be differentiated. First there is the probability that some unfortunate chance event will arise, and secondly there is the magnitude of the consequence. Sometimes use is made of the 'expected value' of a risk, and this is calculated by multiplying the probability of an event by the cost of its consequences. The use of the word 'value' in this context does not imply that it has a positive worth, but simply that it is a measure in terms of money. Sometimes consideration of the expected value of a chance situation can help decide how to handle it; for example during the construction of bridge foundations there may be a 1 in 20 chance of a river flood above a specified level, and this would cause damage with an estimated cost of £1 million. The 'expected value' of the damage would then be £1 million divided by 20, i.e. £50 000. If a protective cofferdam would cost less than £50 000 then it would seem advisable to build one, but if the cost was greater than this sum it would be open to doubt. There may of course be other factors to take into account, such as delay to the project, bad public image and so on, but it does provide an indication of what should be considered. The whole subject of risk management is discussed further in later sections of this chapter and elsewhere (Edwards, 1995).

Consequences of failure

In the preceding section risk was calculated only in financial terms, but there will often be other considerations to take into account. Where life is put at risk it will be necessary to look at other measurements, and in cases where the environment may suffer, as in the case of toxic leakage, it is not appropriate to think in terms of money alone. In an attempt to clarify the earlier definitions of certainty, uncertainty, variability and risk it may be helpful to look at two examples of projects in the construction industry.

The first example is that of the construction of a concrete water tower supplying a village. First the structural design must be examined; once the capacity of the tank has been decided it will be possible to determine the self-weight of the structure and the loading due to water with a fairly high degree of certainty. The dimensions of the tank will only vary from the drawings by a few millimetres, and the density of concrete is within a tight range. It is not possible to overload a tank with water since it will simply overflow and therefore all the static loads on the tank are known with confidence. The dynamic loads are quite different; wind speeds will certainly reach moderate levels, but very high wind speeds will be less likely, and the problem becomes one of deciding the figure to choose. The problem is exacerbated by the fact that the wind load is proportional to the cube of the wind speed, which means that a doubling of wind speed brings about an eight-fold increase in wind load; it therefore quickly becomes prohibitively expensive to 'add a bit for luck' when deciding on the speed to select for the design. Many other factors must be considered, among them the planned life of the structure and the consequence of failure. Given that such tanks are usually at the top of a hill they are inevitably exposed, but at the same time it is highly unlikely that anyone will live or work close to them and be in real danger. There is also the matter of loss of water supply in the area, but this may be considered to be serious but not catastrophic. It is therefore likely that in this design there is a high degree of certainty about all loadings except wind, and that even that would not be a disastrous threat.

A further consideration in some parts of the world is that of earthquake. In most areas of the UK the possibility of an earthquake of significantly dangerous magnitude would not be considered in the design of a water tank of this type. In the case of structures where the consequence of failure is more serious it is necessary to evaluate earthquake loadings; this has certainly been done in the design of many nuclear installations. Compare now the water tank with the next example.

This concerns the construction of an office block over a railway station. Here many of the factors are different. The self-weight of the structure will be accurately predictable, but the loading on the floors may in this case not be easy to determine with confidence. It is generally the case that office buildings have fairly light loadings, and advantage of this fact is taken in the design. However, there is always the possibility that a building of this type may undergo a change of use at some stage, and if it should be used as a warehouse the loading could increase many times over. If this were to be allowed for in the design, the cost would be very much higher, and therefore it is usual to impose limits to floor loading, but this may be difficult to enforce. Wind-loading in downtown areas is less severe than on hill-tops but it must be considered; however more importantly this building does present a hazard to a large number of people due to the fact that it is over a railway station. Because of this last fact it is important to keep the risk low, and this can only be achieved by keeping the probability of failure very low, because consequential loss would be high.

This discussion of variability, uncertainty and risk has mostly been in the context of the decision-making phase of a project, and it is now important to see how these three factors are handled in terms of the management of the execution stage. The first question to look at is the time estimates for the completion of activities.

VARIABILITY OF TIME ESTIMATES IN PROJECT NETWORKS

This topic presents a dilemma. In Chapter 5 detailed methods were described to calculate project intermediate and completion dates, and to use these as a basis of project control. These calculations are mostly simple arithmetic but are tedious; they are also very precise. Given a logic diagram and a set of activity durations there is only one 'right answer' for the calculated project duration, but really it is known that many of those activity durations were only rough guesses in the first place. What has happened is that a series of very precise calculations has been carried out using somewhat imprecise data; this might be analogous to setting out a building by visual inspection of the drawings and then pacing it out on the ground. What we really know is that for any given activity different people would make different time estimates, and different squads of operatives would take different times to do the work. It would be more realistic therefore to quote not a single time estimate for each activity but a range within which it is likely to fall. It is possible to do better than this and estimate a probability distribution of times required for each activity. Unfortunately this then opens the door to much more complicated calculations, and many project managers do not feel that it is worth the effort. At the same time researchers in project management feel that the more complex methods they have evolved present the only true way to model the behaviour of real projects. With this dilemma in mind the following sections look at how the variability of times can be handled, and give an indication of the impact of such variations which can in some projects be very significant. A more thorough description of the methods used is left to other books which deal with the subject in considerable depth (Moder et al., 1983; Spinner, 1992). These books deal with Programme Evaluation and Review Technique (PERT) and probabilistic networks, where not necessarily all activities are completed, e.g in the case of research and development projects. The latter present a very specific case of projects, since some of them will never be complete; others may re-cycle around a series of repeated trials until success is achieved, and many are subject to very large variations in activity durations because of their exploratory nature. Such projects were referred to as 'open projects' earlier in this book, and this title does give an idea of their nature, as many development projects are really never-ending. These are not discussed further here, but anyone facing the management of development projects should refer to the specialised texts. Prior to consulting these texts it would be useful to become familiar with basic statistics including probability distribution and combination of probabilities (Hamburg, 1974).

Variation of time estimates for activity durations

The PERT approach to this is to state three estimated times for each activity

a = optimistic time (less than this on 10% of occasions)
b = most likely (50% chance of completion within this time)
c = pessimistic time (more than this on 10% of occasions).

The first calculation then is to calculate an expected value of the time estimate using the expression

$$\text{expected duration} = (a + 4b + c)/6$$

In the case where $a = 4, b = 6, c = 10$ this would give a value of $(4 + 24 + 10)/6$, i.e. 6·333. This is really little different from the value of exactly 6 which was estimated as the most likely in the first place, and it hardly seems worth the effort of carrying out the calculation; if this is the only use that is made of the three time estimates then it is certainly of limited value. Rather than making use only of the weighted mean of time estimates, it is much more meaningful to look at the range of values and see how these interact, one activity with another. Consider now the very simple network of Fig. 60 in which the three time estimates a, b, c are marked below the activity arrows in brackets. Using the earlier formula, the expected duration of activities 1–2 and 1–3 would be

$$\text{for } 1\text{--}2 \quad (2 + 4 \times 4 + 6)/6 = 4$$
$$\text{for } 1\text{--}3 \quad (3 + 4 \times 4 + 5)/6 = 4$$

and this would indicate that the expected time of completion of the project would be at time 4, if we only make use of the expected duration of each activity. However there is really only a 50% chance that each of the two activities will be completed within 4 days, and for the project to be complete it must be remembered that both of these activities must be complete. Here it is necessary to combine the two probabilities by multiplying them together

$$\text{(probability of completing } 1\text{--}2) \times \text{(probability of completing } 1\text{--}3)$$
$$\text{i.e. } 0{\cdot}5 \times 0{\cdot}5 = 0{\cdot}25$$

There is therefore only a 25% chance that the project can be completed in 4 days, rather a different impression from the one given earlier. In Table 11 the approximate probability of completing each activity within the time shown is given, and hence by multiplying them in pairs the chance of completing the project within that time is shown in the last column.

From this tabulation it can be seen that there is only a 25% chance of completing in 4 days, with 6 days being more likely and 7 a safer prediction. It is not suggested that this full analysis should be carried out for every activity of every network, but it does demonstrate the fact that converging jobs can interact with the potential to

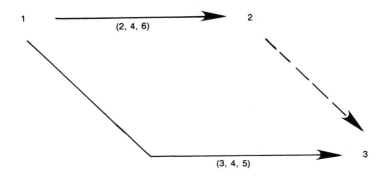

Fig. 60. Three time estimates for each activity, used to calculate the chance of completing on time, and showing the problem of convergence

Table 11. The probability of completing project within stated time

Time: days	Probability of each activity being complete by time stated in left-hand column		Probability of event 3 being achieved by time in left-hand column
	Activity 1–2	Activity 1–3	
2	0·1	0	0·1 × 0 = 0
3	0·25	0·1	0·25 × 0·1 = 0·025
4	0·5	0·5	0·5 × 0·5 = 0·25
5	0·75	0·9	0·75 × 0·9 = 0·675
6	0·9	0·99	0·9 × 0·99 = 0·89
7	0·98	1·0	0·98 × 1·0 = 0·98
8	1·0	1·0	1·0 × 1·0 = 1·0

cause a delay to the project; all activities which converge at an event must be complete for the project to proceed. It is at least necessary to consider this possibility and make allowances in time estimates to guard against it, but there is another way of tackling the convergence problem, as follows in Fig. 61.

First a calculation is described using a simplified but still rather detailed statistical approach. It may be that some readers will feel that this is unnecessarily complex and will prefer to miss out this section of the text. An understanding of the previous paragraph may suffice for an appreciation of the impact of variations in time estimates. For those who already have a statistical knowledge the following will be simple to understand. The network shown in Fig. 61 has marked on each of the activities three time estimates, as previously described. It would be fairly simple in this case to calculate the expected duration of each of the activities, and use these to find an expected duration for the project as a whole; this has been done in Fig. 62 and Table 12 which follows. Using the methods described in Chapter 5 the earliest event

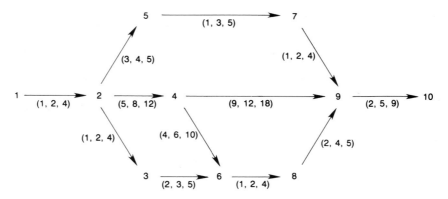

Fig. 61. Three time estimates for activities in network described

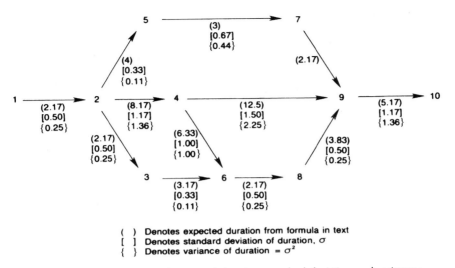

( ) Denotes expected duration from formula in text
[ ] Denotes standard deviation of duration, σ
{ } Denotes variance of duration = σ^2

Fig. 62. Results of calculations of expected duration, standard deviation and variance

times (EET) and latest event times (LET) are calculated, with the expected durations of activities as shown in Fig. 62.

The network diagram shows the expected duration of each activity marked below each arrow, e.g. (2·17), but also has marked on each arrow the 'variance' for each activity duration, e.g. {0·25}. The variance is a statistical measure of the spread of the distribution of time estimates for each activity, and is equal to the square of the standard deviation. For simplicity it is possible here to take the standard deviation as one-sixth of the range. In the case of activity 1–2 the range is from 1 to 4 days, and hence the standard deviation is 0·50 and the variance 0·25. The values of variance are marked on Fig. 62, e.g. {0·25}. There is a statistical relationship which states that when a series of distributions is added together, then the variance of the sum is equal

Table 12. Expected duration of project as a whole

Event number	EET	LET	Note
1	0·00	0·00	critical event
2	2·17	2·17	critical event
3	4·33	13·50	
4	10·33	10·33	critical event
5	6·17	17·67	
6	14·50	16·67	
7	9·17	20·67	
8	16·67	18·33	
9	22·83	22·83	critical event
10	28·00	28·00	critical event

to the sum of the variances. This is what we are doing when we add together the durations along the critical path. If we now do this for Fig. 62 we find that the sum of the variances is $0.25 + 1.36 + 2.25 + 1.36 = 5.22$, and the square root of this, 2.28, gives the standard deviation of the total duration of the project. Statistical tables tell us that there is a probability of 0.50 that the project will be completed within the expected duration, 0.84 probability that it will be one standard deviation more than the expected duration and 0.94 that it will be within two standard deviations. With the numbers already calculated this means that the following will apply

0.50 chance of completion within 28.00 days

0.84 chance of completion within 30.28 days

0.94 chance of completion within 32.56 days

Convergence

The above times can however be misleading because of convergence of the form illustrated in the discussion of Fig. 60. Convergence of activities does however present a problem as was seen in the earlier example of Fig. 60. Examination of the network in Fig. 61 shows that there are two events at which activities converge, namely 6 and 9. The longest possible time to complete the path 2, 3, 6 is $4 + 5 = 9$ days, and the shortest possible time to complete the path 2, 4, 6 is $5 + 4 = 9$ days, and therefore it is impossible for path 2, 3, 6 to determine the earliest event time of event 6 (since all possible values of the duration 2, 3, 6 are equal to or less than the lowest possible value of the duration 2, 4, 6). Convergence at this event is therefore not a problem. Turn now to event 9 where three paths converge and compare the range of values of the durations of those three paths as shown in Table 13.

From the figures in Table 13 it can be seen that the range of values of duration of path X does not overlap at all with the range for path Y and therefore path X will never determine the earliest event time of event 9. The overlap between paths X and Z is so small that it can be ignored for all practical purposes, but there is clearly an overlap between paths Y and Z. Remembering that both paths Y and Z must be complete the procedure now is to determine what the probability distribution is for the earliest event time of event 9. The calculation of this is somewhat tedious, but it is clear that the previous simple calculation that gave a value of 28 days for the path 1, 2, 4, 9, 10 could be misleading since it ignores the possibility that activity 8, 9 could influence that time. The expected event time for event 9 by that first calculation is 22.83 days and for the completion of activity 8, 9 it is 22.67 days with

Table 13. Three paths — comparison of range of values

Path	Events	Minimum duration	Maximum duration
X	2, 5, 7, 9	$3 + 1 + 1 = 5$	$5 + 5 + 4 = 14$
Y	2, 4, 9	$5 + 9 = 14$	$12 + 18 = 30$
Z	2, 4, 6, 8, 9	$5 + 4 + 1 + 3 = 13$	$12 + 10 + 4 + 5 = 31$

a standard deviation of 1·22. The statistical tables would then show that there is 0·54 probability that 8, 9 is completed before 4, 9, and this means that there is a good chance (0·46) that 8, 9 will not be completed until after 4, 9, and will therefore extend the expected duration of the project.

Simulation

All of this calculation by hand is very daunting, and fortunately it can be carried out by computer, but even then requires much data, involving every activity in the network. There is a simpler approach, simulation, which will carry out the analysis satisfactorily, also using a computer. The procedure for this is to assess a time distribution for each activity and then by using a random number generator select for each activity one value of duration. The earliest and latest event times for the network are then calculated, and this represents one possible outcome for the project. The procedure is then run for several iterations, and from this a distribution for each event time will be produced. It will also become clear whether there is one unique critical path or perhaps some other paths which become critical in a few of the simulation runs. This method may seem to be approximate, but within the accuracy of the initial time estimates it does give acceptable answers. A relatively simple exercise can be carried out to illustrate this method using a pair of dice as a random number generator, and carrying out the time analysis by hand.

Simulation exercise using dice   If two dice are thrown the sum of their spots will be in the range 2 to 12. There is only one way to achieve a throw of 2, namely two 1s. There are two ways to throw a total of 3, namely a 1 and a 2 or a 2 and a 1. Table 14 shows the total pattern.

The triangular distribution, while not very close to the 'normal' distribution used by statisticians, gives a rough approximation of what may happen in a project activity. The next step is to tabulate the three time estimates for each activity in the project and assign time values for each activity for each dice throw sum, as is set out in Table 15 (which relates to the network of Fig. 61).

Table 14. Simulation exercise using dice

Total	Combinations of two dice	Number of combinations
2	1, 1	1
3	1, 2; 2, 1	2
4	1, 3; 2, 2; 3, 1	3
5	1, 4; 2, 3; 3, 2; 4, 1	4
6	1, 5; 2, 4; 3, 3; 4, 2; 5, 1	5
7	1, 6; 2, 5; 3, 4; 4, 3; 5, 2; 6, 1	6
8	2, 6; 3, 5; 4, 4; 5, 3; 6, 2	5
9	3, 6; 4, 5; 5, 4; 6, 3	4
10	4, 6; 5, 6; 6, 4	3
11	5, 6; 6, 5	2
12	6, 6	1

Table 15. Three time estimates for each activity using simulation

Activity	Three time estimates			Dice throw sum											
	a	b	c	2	3	4	5	6	7	8	9	10	11	12
1, 2	1	2	4	1	1	2	2	2	2	3	3	3	4	4
2, 3	1	2	4	1	1	2	2	2	2	3	3	3	4	4
2, 4	5	8	12	5	5	6	7	8	8	9	9	10	11	12
2, 5	3	4	5	3	3	3	4	4	4	4	4	5	5	5
3, 6	2	3	5	2	2	2	3	3	3	4	4	5	5	6
4, 6	4	6	10	4	4	5	5	6	6	7	7	8	9	10
4, 9	9	12	18	9	9	10	11	12	12	13	14	15	16	18
5, 7	1	3	5	1	1	2	2	3	3	3	4	4	4	5
6, 8	1	2	4	1	1	2	2	2	2	3	3	3	4	4
7, 9	1	2	4	1	1	2	2	2	2	3	3	3	4	4
8, 9	2	4	5	2	2	2	3	3	4	4	4	4	5	5
9, 10	2	5	9	2	2	3	4	5	5	6	6	7	8	10

Now throw the dice once for each activity and read off from Table 15 the duration to be used for that activity. Repeat the dice throws so that you will end up with two simulations for the project.

These durations are then entered into the same network and the event times calculated as in Fig. 63 for the first simulation and Fig. 64 for the second. In both cases it is seen that the critical path passes through events 1, 2, 4, 6, 8, 9, 10, and it

Table 16. The duration to be used for each activity calculated using simulation

Activity	First simulation			Second simulation		
	Dice	Sum	Duration	Dice	Sum	Duration
1, 2	3, 2	5	2	4, 5	9	3
2, 3	1, 6	7	2	6, 5	11	4
2, 4	6, 5	11	11	6, 1	7	8
2, 5	4, 2	6	4	2, 3	5	4
3, 6	3, 2	5	3	4, 4	8	4
4, 6	4, 2	6	6	1, 4	5	5
4, 9	2, 1	3	10	5, 2	7	12
5, 7	1, 5	6	3	1, 2	3	1
6, 8	3, 5	8	3	1, 6	7	2
7, 9	1, 3	4	2	3, 5	8	3
8, 9	4, 5	9	4	6, 1	7	4
9, 10	5, 3	8	6	1, 6	7	5

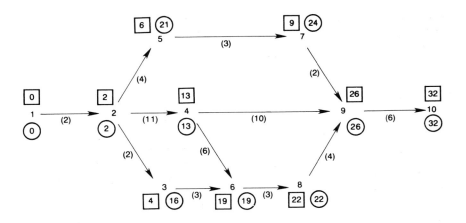

Fig. 63. First simulation of network project duration

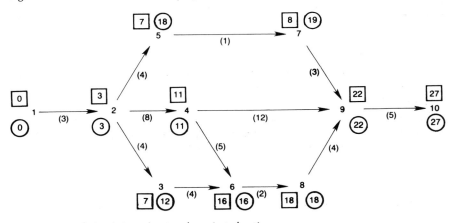

Fig. 64. Second simulation of network project duration

has a total length of 32 days in the first case, but only 27 in the second. These compare with a duration of 28 days calculated for a slightly different path 1, 2, 4, 9, 10 using the PERT method in Fig. 62. Many further iterations should be made, and it will be found that the total project duration will have a distribution, from which it is possible to deduce the probability of meeting a particular target date. Any readers who have the time and interest can use this method to calculate further iterations of this simulation, and hence plot a probability distribution for the duration of the project as a whole. In reality the whole process can be carried out quickly by using a computer-based simulation package.

It is most unlikely that project managers will go through all of these calculations or simulations for a typical project where there are often long sequences of activities in paths which do not interact with each other in very many places. It is however of value and interest to go through this sort of analysis in the case of a project where there are many cross links between paths, especially if they are of similar duration;

these are the cases where such interactions can very quickly lead to overall delays in projects. It is worth trying this type of analysis so that the highly interactive cases can be recognised and dealt with appropriately.

RISK MANAGEMENT

This topic was referred to earlier in this chapter, but most of the content of the subsequent notes concerned the variability in time required to complete activities, with the ways of handling this in network calculations. The essential uniqueness of projects, with the absence of exact prior experience, means that there is a high chance that the expected outcome will not occur exactly. In other words there is a risk that either in terms of performance of the finished product, or its cost or time targets, there may well be a deviation from the plan. In principle there is a chance that this deviation could be in a positive or negative direction, but in practice the maximum possible benefit is much smaller than the maximum possible loss (e.g. if a job is expected to cost £100 the best possible outcome is that its cost is zero, a saving of £100; it could however go badly wrong and cost £1000, an increase of £900). One interesting aspect of risk is the human attitude to chance, especially remote chance. Virtually everyone buying a national lottery ticket does so in the hope of winning the jackpot, and thinks that it could happen. There is perhaps a higher chance that they will be involved in a serious accident on their way to the shop to buy the ticket, but this is dismissed as something that will happen to someone else. Unfortunately it does seem to be the events with undesirable consequences that arise most frequently in projects, and therefore our concern in project management is that things may go wrong, i.e. there is a risk that the project will be below the targets of performance, cost or time. We should seek to minimise the impact of these risks, or at least manage them; this is the subject of risk management (Thompson and Perry, 1992).

Two main types of risk occur in projects

(a) variation from expectation of time, cost or performance
(b) a chance event, e.g. a flood.

In the first case there is a high probability of small losses, and in the second a low probability of high losses. The two are quite different and have to be handled appropriately.

The successive principle in estimating

Consider the case of a building services company with a contract for a total services installation in an existing factory building. There is reasonable confidence in the cost of materials and components as these are standard items, with little risk if they are locally produced items. If however much of the plant is imported there is the risk of exchange rate escalation, and this should be guarded against by seeking fixed local prices, or by forward buying. The labour cost of fixing may be subject to slight uncertainty and this can be largely avoided by the use of appropriate payment schemes. However a third area of the work, the cutting of openings in the existing building may be subject to much greater uncertainty, especially in the case of an old fragile structure; perhaps the best way to avoid risk here is to sub-contract the work at a fixed price to someone with more appropriate experience and equipment.

This approach to managing the risk resulting from variability in cost is to break the project down into a number of jobs which offer more accurate estimates. These can in turn be broken down into smaller elements, with consequent greater accuracy. This approach was evolved by S. Lichtenberg (Lichtenberg, 1984), and is founded upon the principles of statistics relating to the summation of means and variances. In non-technical terms the basic idea is that if many elements are involved there is a high probability that they will not all vary in the same direction, and the concept of swings and roundabouts will apply. There is also a better opportunity to isolate the major potential risks and to deal with them specifically and in isolation. Another aspect of this approach is that it is often found that risk can be reduced by seeking further information, in this case breaking the task down into smaller elements will of necessity give a greater understanding of what has to be done and what potential problems might arise. This concept of seeking further information will be found in many areas of the project.

Chance events

The chance event, e.g. the impact of an exceptionally high tide causing flooding of the project site is one which can be managed in one of two ways. One approach would be to spend money on a flood embankment to keep the water out. The other is to take out insurance, to meet the cost of damage and repair or replacement. Again the idea of seeking further information may be valid, perhaps simply by obtaining more detailed or more up-to-date information on the potential for flood. It may be useful to find out if there have been any major local river works in recent years which could influence flooding for better or worse. It is well known that extensive building development leads to the more rapid run-off of water from an area, potentially causing flash floods, and there have been many examples of this. If it is recognised that there is a serious risk but not enough local flood information is available it may be advisable to carry out hydraulic studies, possibly including model testing — again an example of buying more and better information.

The ability to take risks

It is the magnitude of the risk and the financial resources of the risk-taker that will determine whether action is needed to avoid risks, not the level of probability that they will occur. For example, a small company could not survive the total loss by fire of a significant building on which it is working, and must therefore ensure that adequate insurance cover is in place. By contrast a large company working on several small projects may elect not to insure them because the total of the insurance premiums is greater than their average losses. (Many large car hire firms will not take out comprehensive motor insurance, and prefer to meet the cost of repairs averaged over a large fleet — they are said to be self-insured.)

The overall approach to the management of a risk is to anticipate what may happen, and then analyse it, and in particular determine its magnitude and assess the probability of it arising within the project duration. Analysis of the risk involves determination of whether it is a variation from plan or a chance event. Variation from plan can involve any of the three main parameters that are controlled in project management, namely time, cost and quality. Chance events will include fire, flood, storm and other natural disasters, accidents, political and social events, technical failures, marketing failures, labour problems and liability to other parties. It could be said that risk management is the approach which has as its aim a zero risk policy; this may at first seem to be a forlorn hope, but in the case of manufacturing the concept of a zero defects policy has been adopted for many years, with considerable success. The ideal of working towards zero risk depends on looking forward to see what risks may arise and then taking action to avoid or at least to minimise them. This is entirely consistent with one of the general principles of project planning, namely to 'look before you leap', rather than the more bullish approach adopted in the past by some managers of projects of 'let's get on with it'.

The sequential approach to risk management
Identification
The procedure is to examine the project plan and to identify areas of potential risk in terms of variations from planned work. Are there areas where it is possible that the project might not attain the required performance standards? Is there a risk that cost performance could be below target in some of the activities? In which of the activities are time estimates open to doubt? In almost every project there will be a considerable number of these. Secondly the project should be looked at from the point of view of random chance events. What are the risks of storm, flood, earthquake and other natural disasters? These are generally recognised as being outside the control of the project team, but the team may be able to minimise their effect. What are the risks of fire, accident and other chance happenings which can to some extent be prevented by project managers?

Assessment
Assessment of the risk then follows, in terms of two parameters

- the probability that a variation or chance event will occur
- the magnitude of the risk, i.e. the consequence of the risk in terms of loss of time, money, life, reputation or other measure.

Chance of variation
The chance of variation in terms of time, cost and performance, can be measured by looking back at past performance in similar work. It is well known that in many project-based industries, especially construction, output in these terms is very variable, but it may take considerable work to actually measure these and it could be felt that the effort is not justified. This is perhaps where consequence of variation

comes into the reckoning; if an activity is on the critical path or within a short time of being so, then the effect of an overrun in time is serious and it is worthwhile looking fully into the time estimate. On the other hand activities with large float values are unlikely to delay the project, and the concept of swings and roundabouts can be invoked. It is however important to be satisfied that all the potential variations are not in the same direction, usually adverse, since if they are the float will soon be used up and the project put at risk. This applies also to cost and quality performance.

Consequence of variation

It is perhaps stating the obvious to say that it is important to examine the effect of variations from planned values of time, cost and quality. Time delays on a project are nearly always serious for both client and supplier/contractor; cost variations will have different impacts on the parties involved, depending upon the form of contract, but someone will have to meet any increased costs which arise. It has been stated in Chapter 7 that acceptance of quality standards which are lower than those specified should never be permitted. This means that any downward variation in quality which takes it below the standard required will result in the rejection of the work and the need for it to be replaced; this is always a serious consequence in projects because it has a knock-on effect on other work. If on the other hand defective work is tolerated it may show up as a more serious fault later.

Total assessment of risk

Any risk that has both high chance of occurrence and high consequence must certainly be further analysed in detail, and some positive action must be taken; any that has either high chance or high consequence should clearly be carefully looked at with a view to action being taken. Even where both chance and consequence are low it is worth further examination simply because the risk was identified in the first place, and also because in such cases it is often cheap and simple to minimise it. There is no totally satisfactory way of combining chance and consequence of risk, but it may sometimes be of interest to calculate an 'expected value' of each risk by multiplying its cost by its probability. For example, if it is thought that the consequence of a fire on a project would be damage of £100 000, and the chance of it arising during the duration of the project is 0·001, then the expected value of that risk would be £100 000 × 0·001 = £100 which is an overall measure of the risk which can then be compared with the cost of taking action to avoid the risk, e.g. an insurance premium.

Managing a risk

There are various ways of dealing with risk, although the insurance industry would like everyone to believe that they have the answer to the problem. Among the actions which can be taken a number are set out below. They can be regarded as either seeking to reduce the risk, transferring it to someone else or reducing its impact.

Variation in performance of time, cost and quality can be reduced by employing more highly-skilled operatives and giving them better training. Incentive schemes can be structured in such a way as to encourage all-round good performance rather than simply encourage economy at the expense of quality. More and better supervision with full and clear working instructions will enhance performance in all respects. Particular care with the elimination of errors will help to eliminate sub-standard work and its resultant costs and delays; these measures all amount to better project control.

Within a project, one way to deal with areas of uncertainty that has already been discussed is to seek more information, possibly at some cost of time and money. Work underground is often a problem in this respect, for example the case of driving a tunnel where by its very nature it is impossible to study all of the subsoil in detail before commencing work. An examination of the two ends of the tunnel will give some limited information, which can be supplemented by examination of geological maps. It is possible that there is significant variation of ground along the length of a tunnel, and if possible it is advisable to take bored samples at a number of points along the line of the tunnel. These may be difficult, e.g under a river, and expensive but will provide better information on the cost and time of construction, and may even affect the route. The amount of information provided by each bore-hole will diminish with each one bored, so at some point the cost of another one would not be justified by the extra information obtained; this will then determine how many test bores should be drilled.

Another way of dealing with the unknown is to transfer the work to someone else, and this is not necessarily a matter of passing the buck. It is often the case that a specialist company has the equipment, skilled staff and experience to carry out specialised work, and will offer to undertake it and accept the responsibility that goes with it. A main contractor may estimate some work and submit a tender price of £100 000, but on further examination sees that there is a possible range of error from £80 000 to £150 000. If it is possible to sub-contract the work to a specialist at a price of £110 000 then the main contractor will remove the uncertainty with its potential loss of £50 000 but will have to pay £10 000 to do so. The decision will have to be balanced against a possible gain of £20 000, and the benefit of not having to carry a large risk.

On some projects problems of quality arise in relation not just to an absolute standard, but in terms of consistency. This can happen for example in the colour of component parts, the appearance of some handcrafted work, and similar characteristics which are not easily defined. Consistency can be helped by employing the same group of people throughout a project, ordering enough component parts for the whole project at one time, or by entering into partnering arrangements with specialists who can offer the consistency of their work.

Chance events can also be handled in various ways. Action can be taken to prevent working accidents, and much recent implementation of legislation in the construction industry has been aimed at this. Prevention of fire can be achieved by specific action in terms of the materials used, the design and location of the project, working practices and proper training and supervision. Flood prevention is probably more a matter of design of both temporary and permanent works, e.g. the building of a cofferdam around a site near a river. All of these measures will reduce the impact of chance events, and most of them will cost money; decisions must therefore be taken on whether the cost is justified by the benefit of eliminating the risk.

Insurance does provide one of the means of transferring the risk to someone else, but what it really does is to average out all the costs to those who suffer from chance events, rather than to eliminate them. It does make risk on any one project bearable, and in some ways can actually diminish risks somewhat. This comes about because the insurance companies naturally wish to reduce the magnitude of the risks they carry by requiring certain standards of accident and fire prevention. They will also frequently discourage working in some areas, e.g. those liable to flooding, by quoting high premiums. Insurance companies have a great deal of experience of risks, and their advice will often lead to both a reduction in risk and a reduction in premium rates for insurance cover.

A zero risk policy is then something that could be envisaged as an objective, and possibly attained by a combination of measures such as better information, more training and supervision, off-loading risk onto specialists, preventive measures and insurance. If any other attitude is adopted it means that some risks are being accepted as unavoidable, a view which cannot in the long run be acceptable to project managers as a profession.

14 Project management information systems

It is very clear by this stage of the book that there is a great deal of information to be handled in the management of projects, and furthermore there are many complex but specific inter-relationships to be taken into account. In addition there are many simple but somewhat tedious arithmetic calculations to be performed quickly and accurately. All of these factors point to the use of computers in processing the information required for the management of projects. It so happened that the spread of the use of computers and the growing interest in project management came about at the same time, but this was not because computers are essential for the application of all of the methods used. It was perhaps partly coincidence, but also the fact that the computer companies themselves saw the need for strong project management and accordingly played an important part in its development.

It may seem somewhat strange to leave such a topic as project management information systems (PMIS) until the last chapter of the book, but there is a reason for this. It is important for newcomers to project management to learn the thought processes that go into the analysis and planning stages of a project and then move on to the implementation. In the early days of the discipline it was taught very much as an application of computing and this was certainly a case of 'the tail wagging the dog'. Software companies in particular saw project management as a great potential market for their work, and there are currently more than 200 commercial packages available, all offering their own particular advantages. It is thought to be important in the understanding of the calculations of event times, float, resource smoothing, costing and valuations that project managers must go through the procedures of the calculations themselves. Only then can proper interpretation be placed on the output from computers.

The use of a computer to calculate event times alone can only be justified if it is likely that there will be modifications to the network, which make it necessary to recalculate the times. If a network is to be calculated only once it will actually take longer to complete the computer input than to carry out the calculations by hand. However if an error is made this may not be found out until both forward and backward runs of the event times are complete, and the whole process will have to be repeated if a hand calculation is used. With a computer it is very easy to rerun with adjusted or updated data, and hence in most practical cases the use of a computer is well justified (Jackson, 1986).

The purpose of this chapter is to indicate the types of calculation that can be executed, the forms of presentation of the information, and the ways in which these can be used to plan and control projects. Readers will have to try to work out for themselves which of these facilities are of interest to them and then seek a package which suits the needs of the project they are about to manage. It is not really possible to comment in detail about even a reasonable number of packages, as these are updated frequently and other new ones appear every month; the whole subject area can become very complicated. Like many computer applications which are generally available the packages are usually capable of much more than most users can cope with; this runs the risk of confusion in the minds of those who are not expert in the use of packages.

TIME ANALYSIS OF NETWORKS

This is perhaps the most obvious task that can be performed well and accurately by computer. Some packages are programmed specifically to analyse arrow diagrams and others precedence diagrams; others are able to take data from either network format. The procedure in the simple packages is to enter into the computer the following information for each activity in the network: the event number where the activity starts; the event number where the activity ends; the estimated duration of the activity. These three values for each activity do in fact define both the logical structure of the network and all its timings, and if a listing of such information is provided it is possible to recreate the network. When the time analysis is activated it will produce a listing of the following information for each activity: earliest start date; latest start date; earliest finish date; latest finish date; total float; free float; independent float (not always given and not discussed here). For convenience it is also usual to print out with each activity some other information, i.e. a brief description of the work, for identification; the location by its start and finish event numbers; a marker against critical activities; a responsibility code, by department or company.

Further manipulations can be added by presenting the output information in a number of different forms. Dates can be simply day numbers (or hours, if that is the time unit used), or can be translated into calendar dates. Many packages do this but require that another input gives information on days per week worked (and/or hours per day), and holidays planned. All this data entry takes time. An alternative form of input and output is used in many of the more recent packages. In these each activity is 'drawn' on to the screen in the form of a bar chart, with links to its predecessors and successors. As each activity is entered the screen shows an updated bar chart of all activities entered thus far, as for example in Fig. 65 which relates again to the Photo Products Ltd case.

COST ANALYSIS

Many of the network software packages offer the facility of providing costing and/or value information. This will include a number of different forms. Cumulative cost and

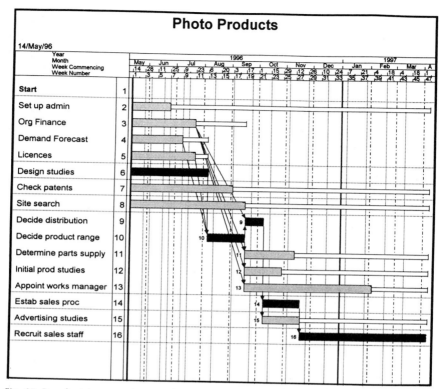

Fig. 65. Bar chart input and output for part of the PP Ltd project (PowerProject Professional, Asta Development, Inc.)

value curves, essentially in the form of S-curves as described in Chapter 6 can easily be produced, but of course they require the input of the relevant data. In Chapter 6 reference was made to 'budget cost', namely the cost that it is expected that the activities of the project will incur. For the analysis of this by computer it is necessary to allocate all the expected costs of a project over the full list of activities, possibly with the addition of an overhead cost based on a daily rate. This task is often quite difficult however because of the form in which the cost estimate has been prepared. If the project consists simply of the purchase of a number of items this may be fairly simple; however if the project is concerned with building or civil engineering it is likely that the estimate has been prepared on the basis of a bill of quantities. In such cases it is most unlikely that the network activities will coincide with the bill items, and it then becomes necessary to make an arbitrary allocation of costs to activities, making sure that total costs are the same and that the allocation is reasonable. This matter can be further complicated in the case of remeasured contracts where the estimated cost is likely to be subject to variation without it being clear which activities are to be modified. Even with these provisos it is useful to be able to print out a 'budget cost of work planned' S-curve to act as the basis of comparison with actual performance.

The first comparison can be made with the 'budget cost of work completed' as the work proceeds, as described in Chapter 6. These two curves are both based on the estimated cost of work which has at the outset been distributed over all activities, and will not change on a day-to-day basis. It will be remembered that this comparison is a means of seeing whether the actual work completed is in total up to programme; it clearly cannot indicate if there are any special problems, or indeed whether progress is being made in the most important areas. It is simply a measure of the total volume of work completed and thus gives a very rough measure of the general state of progress. What it can do is to indicate when there is an overall slip in the project and not enough work is being done in total; the computer package can then predict what is likely to happen to the project duration if work continues to slip at a similar rate.

A second comparison which can be made is between the 'budget cost of work completed' and the 'actual cost of work completed'. This gives a measure of the actual cost of carrying out the work finished and comparing it with the planned or estimated cost. It is therefore a cost control tool, which will require the input of actual cost data. Again this information is needed on an activity-by-activity basis, which is not usually the form in which it will be originally generated. Labour costs will be derived from wages sheets, materials and components from invoices; both of these will need much interpretation to correctly allocate costs to activities, and many project managers will feel that the effort is not justified. This will be particularly true where the nature of the work is changing rapidly and historical costs quickly become outdated. Cost control of individual items therefore becomes very difficult by computer.

Project value does present an area where good use can be made of the computer. At the outset of a project contract it is possible to allocate the priced items of a bill of quantities over all the activities in the network. These values of work, as distinct from their costs, will not change except where the quantities change. At any intermediate stage of the project it is then simple to print out the bill value of work undertaken to date by relating it to the activities completed. This will give the equivalent of an interim valuation for the project and can be used as the basis of progress payments. On one project this procedure was followed in tandem with a conventional interim measurement made by surveyors, and over the period of the contract it was found that the two values agreed within 2%, which is probably about the limit of accuracy of an interim valuation. The ease and speed with which it is possible to prepare interim payment certificates in this way is clearly beneficial.

RESOURCES

It is in connection with resources that computer packages offer great advantages, because of the large number of calculations and manipulations that have to be carried out. Chapter 9 discussed in some detail the nature of project resources and the ways in which they can be managed. The examples in that chapter served to show that the calculations could be tedious in a simple project, and almost impossible in a complex one. This is a situation where a computer package can be very helpful regarding the two separate stages of resource aggregation and resource smoothing.

Resource aggregation

This is carried out initially by summing the number of units of each designated resource for each unit of time of the project; this is usually done on the assumption that each activity starts on its earliest start date. The form of output is a histogram for each resource category; the typical shape of the histogram shows a rapid variation in resource demand, as for example in Fig. 52, often with periods of zero demand included. Chapter 9 referred to 'time-based' resources for which it is usual to plan a smooth demand, since their availability and use cannot be easily varied; hence we seek to plan our projects around the assumption that we seek a steady demand for all time-based resources, and we refer to this as resource smoothing or resource levelling.

Resource smoothing

Resource smoothing is the process of trying to plan work by moving some activities back, within their float periods, in order to create a smooth demand for each resource. This can be done by a software package which includes the appropriate facility. Note that some packages simply state that they have a 'resource facility', and this should be checked to see whether it means smoothing or just the much more simple resource aggregation described above. Resource smoothing is really quite complex, especially if several different resource types are to be used in the project. Care has to be taken about prioritising the resources to be smoothed; consider the following. If the first resource is considered some activities will be delayed within their float periods, which reduces the scope for smoothing the next resource, and so on. Some packages claim to be able to handle up to one hundred resource types, but this only means that there is space within the program to handle one hundred sets of data. In most projects once six or seven resources have been smoothed there is very little float left for further adjustment, unless they refer to totally separate areas of the project. In fact with some packages it is possible that the levelling of a third or fourth resource will upset the earlier smoothing of the first or second. The solution to this problem is to restrict resource smoothing to a small number of the most important resources, e.g. fixed cranes, and it may well be found that sequencing them will automatically smooth out the demand for subsequent resources such as labour skills, which therefore do not have to be processed through the computer.

Variation in time estimates

Chapter 13 described the use of Programme Evaluation and Review Technique (PERT) and the way in which it dealt with uncertainty of activity durations by using three time estimates. This is another application where the tedious arithmetic can easily be handled by computer, but of course it must be remembered that the data of three estimates has to be generated and entered for each activity. Chapter 13 also discussed simulation, and this is the ideal application for a computer, where many iterations of the same type of calculation are carried out on the same set of data. The hand simulation described in Chapter 13 is not realistic for a live project, and was only included in order to explain the method. It is very simple for the computer to make a hundred or more iterations of the calculation and to produce a probability distribution for the event times of all key or 'milestone' events in the project. Such a simulation package is distinct from the usual time and resource analysis packages and is much less widely used.

Multi-project scheduling

This is an application where an organisation is concurrently running several projects. It may well be the case that resources do not need to be smoothed out for each project, but for the company's operations as a whole. This is because a common resource, such as a design team, may work on several projects and can easily switch from one to another as needs arise; in this case there would be no transfer cost as there would be for heavy equipment. Managing a number of linked projects in this way allows a great deal more flexibility in the planning and use of resources. Such planning is however unthinkable without the help of a computer to manipulate the vast amount of data involved.

Development projects

Chapter 13 also considered the type of project where not all activities are completed only once, or perhaps not at all, as might be the case of a research project where several attempts may have to be made to move on to the next stage, and sometimes other steps may be omitted or others added. There may also be alternative paths to follow, depending upon the outcome of uncertain events. This is a very special type of project, and requires a rather different approach which is not within the scope of this book. It is well covered however by Moder et al. (1983).

Integrated management information systems

Several integrated systems are now available which are able to integrate a whole range of the operations of an organisation, possibly linking a project for a new production facility with the operation of an existing plant, the anticipated market demand for product, together with the cash flow for the organisation as a whole, and perhaps even linking it to the anticipated share price of the company. All of this is beyond consideration here, but it helps put projects in an appropriate context.

Appendix 1
Exercises in network logic and time calculations

This series of exercises is provided to give those who are not conversant with network calculations an opportunity to become familiar with them. Either arrow or precedence diagrams may be used, and of course the numerical answers will be the same in both cases.

One of the important lessons to learn is that in drawing the logic network it is vital to be very strict about the sequences, and only show those that are absolutely specific, to the exclusion of those which might be deemed just convenient or usual practice. The examples are deliberately aimed at this point, and for this reason the activities are mostly referred to by a code letter rather than a full description. This is not a practice which should be followed in a real project, where each activity should be clearly described.

Solutions to the exercises are given at the end of this appendix. In all of these other than exercise 4 the logic is quite specific and there is only one right answer, but the diagram may take the form of one of a range of actual visual layouts; provided the logic is correct and the time values agree, any one of them may be correct. It should also be noted that there is no unique solution to the event numbering, one version has been given in the solutions as an illustration.

EXERCISE 1

Draw a logic network in either arrow or precedence format for the following activities.

Activity A	is the start of the project and precedes all others
Activity C	follows activity B
Activities F and G	can start when activity E is complete
Activity D	cannot start before C and F are complete, and must be finished before J can start
Activity J	depends upon the completion of H before it can start

EXERCISE 2

Allocate event numbers to the network shown in Fig. 66, calculate and insert on the network the earliest and latest event times. Then tabulate the earliest start date, earliest finish date, latest start date, latest finish date, total float and free float for every activity in the network.

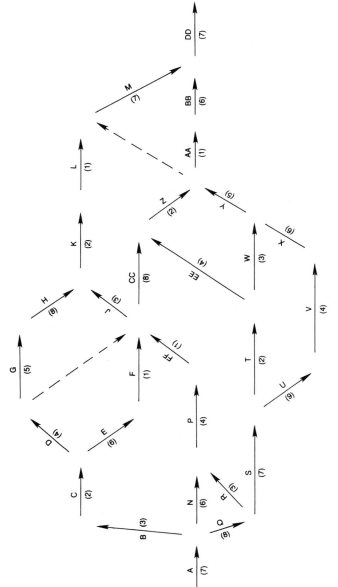

Fig. 66. Exercise 2

EXERCISE 3

Allocate event numbers to the network shown in Fig. 67, calculate and insert on the network the earliest and latest event times. Then tabulate the earliest start date, earliest finish date, latest start date, latest finish date, total float and free float for every activity in the network.

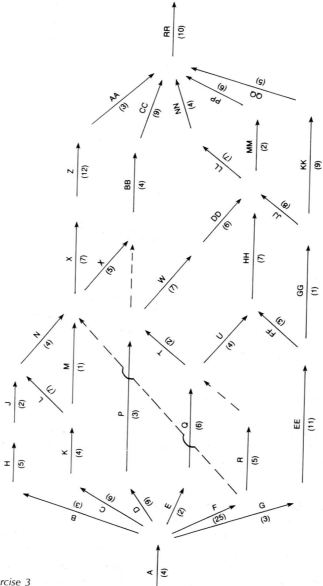

Fig. 67. Exercise 3

EXERCISE 4

Figure 68 shows the layout of a sports ground which is to be built.

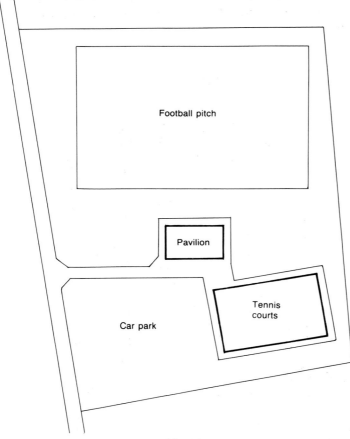

Fig. 68. Exercise 4 — sports ground layout

Part 1

Draw an arrow network for this project, based on the following list of activities. Allocate event numbers and calculate earliest and latest event times, and hence find the shortest likely duration of the project. Exert your own judgement on the necessary logic, but take account of the restrictions below.

- For security reasons the pavilion should not be erected until the fence and gates are complete.
- The pitch and courts will be marked out by the groundsman who will not be appointed until the flat in the pavilion is ready for occupation.
- No money is to be committed until the club committee has approved the plans and estimates at a special meeting to be called for that purpose.

Table 17. List of activities for exercise 4

Duration in days	Activity description
10	Prepare plans
10	Obtain estimate for ground works
5	General site levelling over area of football pitch and car park
10	Construct drains to football pitch
8	Construct drains to car park and tennis courts
6	Construct drains to pavilion
4	Fell trees and remove roots from tennis court area
12	Obtain estimate for prefabricated pavilion building
12	Erect boundary fence
15	Order special entrance gates
3	Erect entrance gates
8	Spread topsoil over football pitch
10	Cultivate and sow grass seed over football pitch
4	Erect football posts
1	Mark out football pitch
3	Lay foundation for tennis courts
15	Lay top surface of tennis courts
4	Erect wire fence around tennis courts
2	Fix markings on tennis courts
6	Build pavilion foundations
20	Erect prefabricated pavilion
8	Lay paths around pavilion and tennis courts
4	Surface car park
2	Connect water from road to pavilion
1	Connect overhead power cable from road to pavilion
6	Decorate pavilion
30	Submit plans for planning permission
15	Order prefabricated building

Part 2

Assume that the project start date is 1 January and that the project times are based on working a 5-day week. Find the impact on the completion date of imposing the following restrictions.

(a) Grass seed may not be sown before May (day 81).
(b) The prefabricated building must be paid for on the day of delivery, but funds will not be available until April (day 60).
(c) It is intended to have 4 weeks of play on the tennis courts before the official opening day so that the final of a tournament can be held on that day.

What is the earliest day on which the opening ceremony can be held? It is understood that the football pitch will not be ready until much later as it will take eighteen months for the grass to mature.

Part 3

It is decided that five of the foundation and drain activities will be carried out by club members on a voluntary basis in order to save money. The number of volunteers is very limited and only one activity can be tackled at any one time. What would be the effect of this resource limitation? The activities affected are

- drains to football pitch
- drains to car park and tennis courts
- drains to pavilion
- foundation for tennis courts
- foundation for pavilion.

EXERCISE 5

Figure 69 sets out a very simple project network, and Table 18 gives information about the activity durations and how much the activities will cost. Calculate the project duration on the basis of normal durations for all activities, and then find the cheapest way of achieving a reduction of 2 days, 3 days and 6 days. Also what would be the lowest total cost of the project if it is carried out in the shortest possible time.

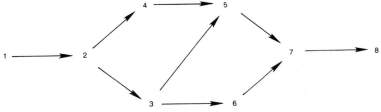

Fig. 69. Network for exercise 5 — time/cost interactions

Table 18. Activities, durations and cost for exercise 5

Activity	Normal duration	Crash duration	Normal cost	Crash cost
1–2	4	3	45	70
2–3	5	3	100	200
2–4	4	3	40	65
3–5	8	4	80	100
3–6	5	3	20	60
4–5	7	3	10	70
5–7	7	5	80	140
6–7	7	4	45	90
7–8	4	2	20	100

245

EXERCISE 6

Construct a network for the project which comprises the following activities and calculate values for each activity of:

- earliest start date
- earliest finish date
- latest start date
- latest finish date
- total float
- free float.

Table 19. Activity durations and dependencies for exercise 6

Activity	Duration	Dependencies
A	7	can start at same time as B, near the beginning of the project
B	4	must precede C, E, M, D
C	12	must precede D
D	6	can be done at the same time as E and L
E	6	precedes L
F	7	follows M and N
G	9	precedes K and L
H	10	follows N
J	9	follows D and L
K	13	follows E and G
L	4	precedes J
M	10	precedes G and H
N	4	precedes H
P	8	follows N and H
Q	11	precedes all other activities
R	6	follows P, F, G, K, L and J

ANSWER 1

The logic network for this problem is shown in both arrow diagram format in Fig. 70 and precedence diagram format in Fig. 71. Note that in both cases only the necessary logic is shown, e.g. no mention is made of any activity following activity G, therefore it can run through to the end of the project. Similarly no mention is made of the start of activity H, except that 'A is the start of the project' and therefore precedes all other activities; hence H can commence as soon as A is complete. Note also that in the arrow diagram the activity description (here just a code letter) is written along the arrow, indicating that the activity can take place over any period between the tail and the head of the arrow. In the case of arrow diagrams the arrow indicates a period of time and the event (or node) is a single point in time. With the precedence diagram these roles are reversed, the activity occupying time takes place at the node (the activity box), and the arrows are simply indicating sequence and occupy no time.

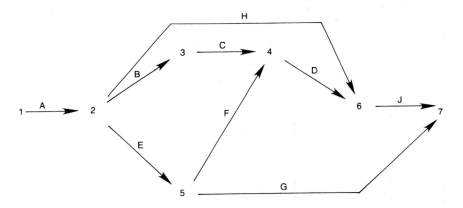

Fig. 70. Logic network in arrow diagram format — answer to exercise 1

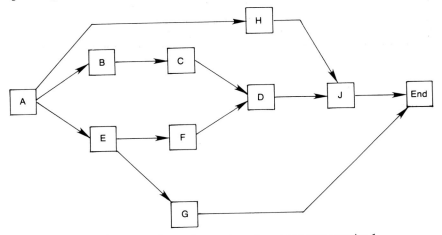

Fig. 71. Logic network in precedence diagram format — answer to exercise 1

247

ANSWER 2

Fig. 72 shows a possible event numbering, but note that there is no unique right answer to that part of the question. The numbers shown are such that each activity has an ascending pair of numbers to identify it; this acts as a check against a loop in the logic, something which is not admissible. The earliest and latest start and finish dates are set out in Table 20.

Note that the start date of an activity is always one day after the corresponding finish date of the preceding activity; this is because work will usually finish at the end of the last day, and the next activity will follow on at the beginning of the next day.

Table 20. Answer to exercise 2

Activity		Duration	Early start	Early finish	Late start	Late finish
A	1–2	7	1	7	1	7
B	2–3	3	8	10	22	24
Q	2–8	8	8	15	8	15
N	2–9	6	8	13	26	31
C	3–4	2	11	12	25	26
D	4–5	4	13	16	27	30
E	4–7	6	13	18	30	35
G	5–6	5	17	21	31	35
H	6–12	8	22	29	36	43
F	7–11	1	19	19	36	36
R	8–9	3	16	18	29	31
S	8–14	7	16	22	16	22
P	9–10	4	19	22	32	35
FF	10–11	1	23	23	36	36
J	11–12	3	24	26	41	43
CC	11–16	8	24	31	37	44
K	12–13	2	30	31	44	45
L	13–22	1	32	32	46	46
T	14–15	2	23	24	37	38
U	14–17	9	23	31	23	31
EE	15–16	4	25	28	41	44
W	15–19	3	25	27	39	41
Z	16–20	2	32	33	45	46
V	17–18	4	32	35	32	35
X	18–19	6	36	41	36	41
Y	19–20	5	42	46	42	46
AA	20–21	1	47	47	47	47
BB	21–23	6	48	53	48	53
M	22–23	7	47	53	47	53
DD	23–24	7	54	60	54	60

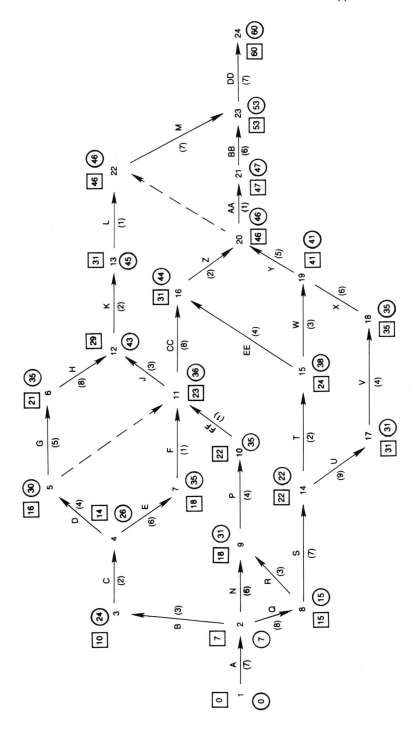

Fig. 72. Answer to exercise 2

ANSWER 3

Fig. 73 shows earliest and latest event numbers and Table 21 sets out the earliest start date (ESD), the earliest finish date (EFD), the latest start date (LSD), the latest finish date (LFD), the total float (TF) and the free float (FF).

Note that on a path where there is float, at the point where that path rejoins the critical path the total float will also appear as free float, e.g. at events 11, 12, 23, and 27.

Table 21. Answer to exercise 3

Activity	Duration	ESD	EFD	LSD	LFD	TF	FF
0–1	4	1	4	1	4	0	0
1–2	3	5	7	25	27	20	0
1–4	6	5	10	18	23	13	0
1–7	9	5	13	25	33	20	0
1–8	2	5	6	27	28	22	0
1–9	25	5	29	5	29	0	0
1–19	3	5	7	26	28	21	0
2–3	5	8	12	28	32	20	0
3–6	2	13	14	33	34	20	7
4–5	4	11	14	24	27	13	0
5–6	7	15	21	28	34	13	0
5–13	1	15	15	38	38	23	14
6–13	4	22	25	35	38	13	4
7–12	3	14	16	34	36	20	20
8–11	6	7	12	29	34	22	22
9–10	5	30	34	30	34	0	0
11–12	2	35	36	35	36	0	0
11–21	4	35	38	39	42	4	0
12–18	7	37	43	37	43	0	0
13–14	7	30	36	39	45	9	0
13–16	5	30	34	43	47	13	2
14–15	12	37	48	46	57	9	0
15–27	3	49	51	58	60	9	9
16–17	4	37	40	48	51	11	0
17–27	9	41	49	52	60	11	11
18–23	6	44	49	44	49	0	0
19–20	11	8	18	29	39	21	0
20–21	3	19	21	40	42	21	17
20–22	1	19	19	41	41	22	0
21–23	7	39	45	43	49	4	4
22–23	8	20	27	42	49	22	22
22–26	9	20	28	47	55	27	0
23–24	7	50	56	50	56	0	0
23–35	2	50	51	53	54	3	0
24–27	4	57	60	57	60	0	0
25–27	6	52	57	55	60	3	3
26–27	5	29	33	56	60	27	27
27–28	10	61	70	61	70	0	0

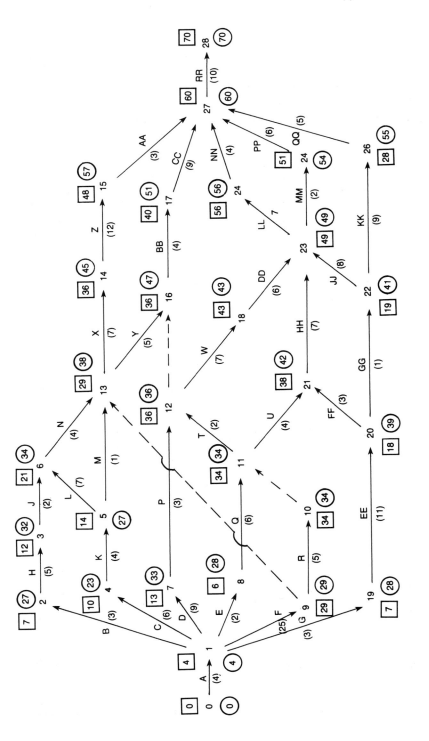

Fig. 73. Answer to exercise 3

ANSWER 4

There is no unique correct answer to this exercise — Fig. 74 gives one possible solution, but it must be emphasised that variations on it are quite feasible. This may at first seem to negate the concept of strict logic, but in practice there will be questions raised about the logic which can only be answered if more information is sought. This in itself is a valuable exercise in project management, since it forces the planner to seek information ahead of the time when it is needed on site; this gives time for the information to be obtained, understood and acted upon. There are a few detailed points to be made about this network.

(a) The requirement that the committee approve all expenditure before any is committed is met by bringing all the estimates and planning together at event number 5. If they are not gathered at one event then it would be possible for one of the estimates to be approved before the other is available; that should not be allowed to occur.

(b) The network is set out in 'bands' of activity on each part of the site, namely the tennis courts, the football pitch, the pavilion and so on. However it should also be noted that some activities span more than one area, e.g. the levelling of the car park and football pitch is carried out as one combined operation, possibly a cut-and-fill which cannot be separated. The drains to the car park are however linked to those in the area of the tennis courts and must be carried out together. These two requirements are covered by inserting a dummy activity 6–9 to ensure that the levelling of the car park is completed before the drains are laid there. Check the logic of your own network to see if this requirement has been met.

(c) The other restrictions set out in part 1 of the question are easily met by ensuring that the boundary fence is completed at event 16 before the pavilion is erected, and by completing the pavilion flat at event 20 before any marking out is done by the groundsman.

(d) Part 2 of the question imposed some further 'restraints' on the project and these are shown together with their impact in Fig. 75. Restraint (a), shown as arrow 2–22, delays the sowing of grass seed. Restraint (b) delays the arrival of the pavilion until cash is available. Restraint (c) ensures that the overall project completion in time for the tournament final is not less than four weeks after the tennis courts are ready for use.

(e) Part 3 of the question relating to resource smoothing will depend very much on the detailed logic of the network. Fig. 76 shows in bar chart form the resources needed for the drain and foundation activities, and shows a sequence in which they could be used for the particular network shown here, and on the assumption that the resource levelling is carried out before restraints of part 2 are considered. The bar chart indicates that there would be a project delay of 4 days. It is not necessary to check all possible sequences, as it can quickly be seen that there is a total of 33 days work to be completed by the gang within a period of only 29 days (days 41 to 79 inclusive), which is 4 days too few. However when the restraints of part 2 are taken into account these would

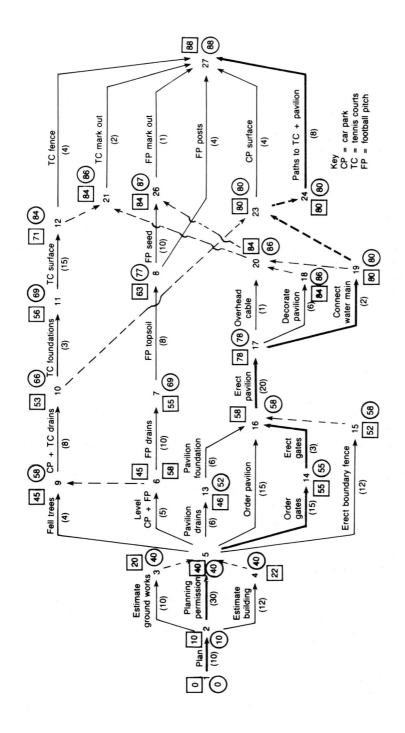

Fig. 74. Answer to exercise 4 part 1 (sports ground)

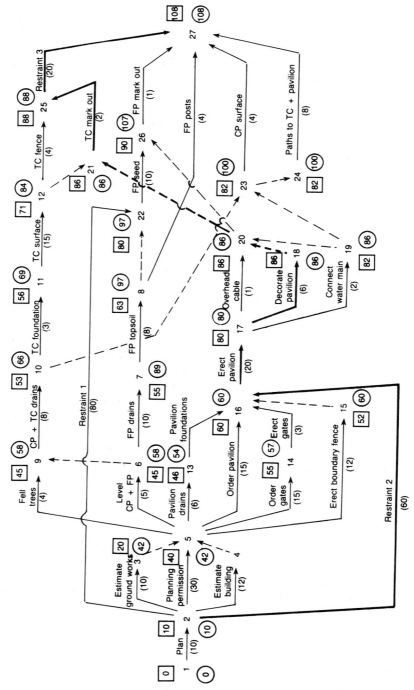

Fig. 75. Answer to exercise 4 part 2 (sports ground)

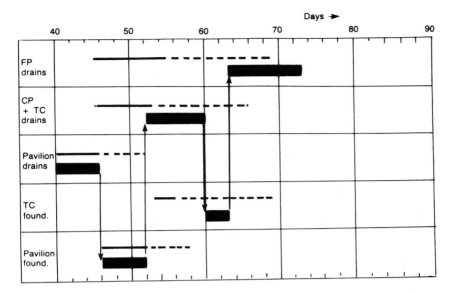

Fig. 76. Answer to exercise 4 part 3 without restraints

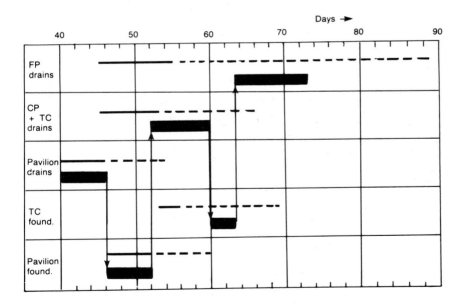

Fig. 77. Answer to exercise 4 part 3 with restraints

themselves lead to an extension of the project duration, and this would give plenty of time for the drain and foundation activities to be completed in sequence with only one gang, as shown in Fig. 77.

ANSWER 5

The first task is to calculate the additional cost incurred by each activity if its duration is reduced by one day. This is done using the expression

$$\text{cost per day saved} = \frac{\text{crash cost} - \text{normal cost}}{\text{normal duration} - \text{crash duration}}$$

For activity 1-2 the additional cost per day is

$$= \frac{70 - 45}{4 - 3}$$
$$= £25$$

The same calculation procedure gives the following cost per day reduction for all activities.

The second step is to calculate the earliest and latest event times using 'normal' durations as shown in Fig. 69. From this it can be seen that the critical path passes through events 1, 2, 3, 5, 7 and 8, and this path must be reduced if the project is to be accelerated. From Table 22 it can be seen that the cheapest way to do this is to reduce activity 3–5 by 1 day at a cost of £5. A second day can be similarly saved, but after that path 2–4–5 is also critical. For a third day's reduction, activities 4–5 and 3–5 would present the cheapest way. This procedure can then be pursued as described in Chapter 8, leading to the result that the additional costs of reducing the project time would be

(a) two days — £10, i.e. two at £5
(b) three days — £30, i.e. three at £5 + 1 at £ 15
(c) six days — £130.

The minimum project duration can be worked out as seventeen days, and the corresponding total project cost (not simply the additional cost) is £830, if it is accelerated in the cheapest way.

Table 22. Answer to exercise 5

Activity	Cost per day reduction: £
1–2	25
2–3	40
2–4	24
3–5	5
3–6	30
4–5	15
5–7	30
6–7	15
7–8	50

ANSWER 6

It is certainly difficult to set out an arrow network for a project in which there is no information given about the nature of the activities, and it may be necessary to make more than one attempt at it. It will perhaps be easier to tackle this example using a precedence format. In fact one of the ways to approach a precedence diagram is to set out a grid of boxes, write the descriptions of activities into as many boxes as there are activities, and then simply connect them by sequencing arrows. Sometimes this leads to a great deal of crossing arrows, but this is still correct logic even though it may look untidy. The two solutions are set out, in Fig. 78 for the precedence diagram, and Fig. 79 for the arrow diagram. It is a useful exercise to try both to see if they agree — they should!

Table 23. Answer to exercise 6

Activity	Duration	ESD	LSD	EFD	LFD	Total float	Free float
A 2–11	7	12	47	18	53	35	35
B 2–3	4	12	12	15	15	0	0
C 3–4	12	16	21	27	32	5	0
D 4–7	6	28	33	33	38	5	5
E 3–6	6	16	29	21	34	13	13
F 8–10	7	26	41	32	47	15	15
G 5–6	9	26	26	34	34	0	0
H 8–9	10	26	30	35	39	4	0
J 7–10	9	39	39	47	47	0	0
K 6–10	13	35	39	47	47	0	0
L 6–7	4	35	35	38	38	0	0
M 3–5	10	16	16	25	25	0	0
N 2–8	4	12	26	15	29	14	10
P 9–10	8	36	40	43	47	4	4
Q 1–2	11	1	1	11	11	0	0
R 10–11	6	48	48	53	53	0	0

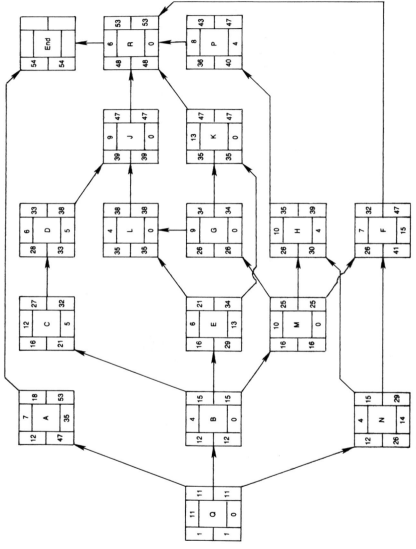

Fig. 78. Exercise 6 — precedence solution

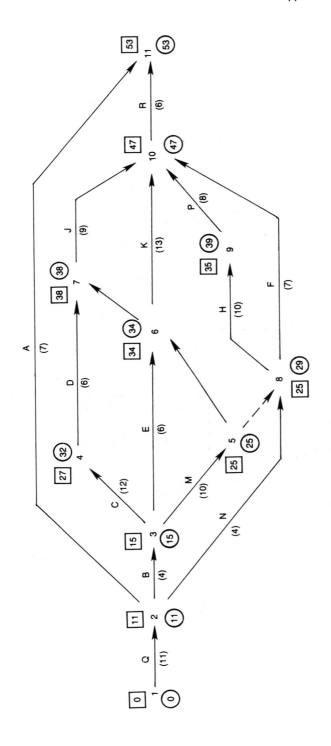

Fig. 79. Exercise 6 — arrow diagram solution

Appendix 2
Table of discount factors

Table 24. Present value of £1 discounted at r per cent for n years

n years from now	\multicolumn{12}{c}{Discount rate r per cent}

n years from now	6	8	10	12	14	16	18	20	25	30	35	40
1	0·943	0·926	0·909	0·893	0·877	0·862	0·847	0·833	0·800	0·769	0·741	0·714
2	0·890	0·857	0·826	0·797	0·769	0·743	0·718	0·694	0·640	0·592	0·549	0·510
3	0·840	0·794	0·751	0·712	0·674	0·641	0·608	0·579	0·512	0·455	0·406	0·364
4	0·792	0·735	0·683	0·636	0·592	0·552	0·516	0·482	0·410	0·350	0·301	0·260
5	0·747	0·681	0·621	0·567	0·519	0·476	0·437	0·402	0·328	0·269	0·233	0·186
6	0·705	0·630	0·564	0·507	0·456	0·410	0·370	0·335	0·262	0·207	0·165	0·133
7	0·665	0·583	0·513	0·452	0·400	0·354	0·314	0·279	0·210	0·159	0·122	0·095
8	0·627	0·540	0·467	0·404	0·351	0·305	0·260	0·233	0·168	0·123	0·091	0·068
9	0·592	0·500	0·424	0·361	0·308	0·263	0·225	0·194	0·134	0·094	0·067	0·048
10	0·558	0·463	0·386	0·322	0·270	0·227	0·191	0·162	0·107	0·073	0·050	0·035
11	0·527	0·429	0·350	0·287	0·237	0·195	0·162	0·135	0·086	0·056	0·037	0·025
12	0·497	0·397	0·319	0·257	0·208	0·168	0·137	0·112	0·069	0·043	0·027	0·018
13	0·469	0·368	0·290	0·229	0·182	0·145	0·116	0·093	0·055	0·033	0·020	0·013
14	0·442	0·340	0·263	0·205	0·160	0·125	0·099	0·078	0·044	0·025	0·015	0·009
15	0·417	0·315	0·239	0·183	0·140	0·108	0·084	0·065	0·035	0·020	0·011	0·006
16	0·394	0·292	0·218	0·163	0·123	0·093	0·071	0·054	0·028	0·015	0·008	0·005
17	0·371	0·270	0·198	0·146	0·108	0·080	0·060	0·045	0·023	0·012	0·006	0·003
18	0·350	0·250	0·180	0·130	0·095	0·069	0·051	0·038	0·018	0·009	0·005	0·002
19	0·331	0·232	0·164	0·116	0·083	0·060	0·043	0·031	0·014	0·007	0·003	0·002
20	0·312	0·215	0·149	0·104	0·073	0·051	0·037	0·026	0·012	0·005	0·002	0·001
25	0·233	0·146	0·092	0·059	0·038	0·024	0·016	0·010	0·004	0·001		
30	0·174	0·099	0·057	0·033	0·020	0·012	0·007	0·004	0·001			
35	0·130	0·068	0·036	0·019	0·006	0·006	0·003	0·002				
40	0·097	0·046	0·022	0·011	0·005	0·003	0·001	0·001				

Appendix 3
Qualifications in project management

There are now several qualifications in project management, which include

- Construction project management NVQ/SVQ (National/Scottish Vocational Qualification) level 5
- Certification by the Association for Project Management (CPM)
- Associated Project Management Professional (APMP).

In each of these cases a candidate's suitability is assessed on the basis of working experience and knowledge, and many courses and texts will be found to be helpful in planning and gaining the appropriate experience.

The levels of knowledge and experience that are appropriate to these qualifications are set out in two documents: *Body of Knowledge*, published by the Association for Project Management (APM, 1996), and *Project Management Skills*, published by the Construction Industry Council (CIC, 1996).

NVQ/SVQ LEVEL 5

This qualification is appropriate for construction professionals in private practice and for those on the staff of client organisations. It is intended for all professionals engaged in construction, and is largely based on the practical experience of having worked in project management in the construction industry. Candidates have to produce appropriate evidence from at least two projects 'of significant size and complexity' on which they have been engaged.

The scope of the work is effectively the full range of the professional aspects of construction. It includes urban, rural and infrastructure development; new build, extension, refurbishment, conservation and alteration. The scope definition excludes the direct management of the work undertaken by contractors and suppliers; reference to the actual construction is limited to those aspects in which the professional team is actively engaged, e.g. evaluating tenders, negotiating contracts, issuing certificates and monitoring quality and progress.

The NVQ/SVQ qualification is primarily concerned with design, costing and formal contractual documents and is not restricted to what is often referred to as the 'project management approach' and the set of techniques associated with it.

The wide range of subjects listed in the NVQ/SVQ documents could not possibly be covered in any course or book, but most are included in the basic professional qualifications.

The 'project management approach', based on the concepts of plan–measure–control, 'right first time' quality, utilisation of resources, variability and uncertainty, risk and project viability, can be applied to all of the areas of construction or other project-based work. These aspects form the underlying theme of this book.

APM QUALIFICATIONS

CPM, the APM's highest qualification, is awarded to senior practising members of the profession who have demonstrated their ability to manage difficult, complex and, in the main, large projects. Registration as a Certificated Project Manager requires completion of a three-part assessment. A candidate submits a self-assessment form accompanied by a précis of a specific project that they intend to present as a demonstration of their project management abilities. Two assessors, one from within the candidate's industry, one from an entirely different field, review these submissions and, if satisfied with the candidate's eligibility, invite them to submit an in-depth report on their project. If this report shows that the candidate has met all the relevant criteria, they are invited to a half-day interview where the candidate's submissions are verified and the breadth and depth of their knowledge, as related to the *Body of Knowledge*, is further explored.

APMP is awarded to individuals actively involved in project management, but in support roles. Its award demonstrates a specified level of knowledge and at least three years' involvement as a leading member of a project team. The two parts of the qualification process are, first, the submission of an application form that allows candidates to demonstrate that they have sufficient experience, and, second, a two-part written examination. The examination involves a two-hour multiple-choice paper to test a candidate's knowledge, and a one-hour written paper to test whether a candidate knows how to apply that knowledge.

Certification is initially valid for three years, but may be extended on an annual basis if the holder can demonstrate a fixed annual amount of continuing professional development. The APMP qualification is retained for life.

TAUGHT COURSES

There are many short courses of various lengths, levels and formats, and these are well advertised in the project management journals. There are now many formal courses, many at masters' level, mostly MSc or MBA, and at least one MPM (Master of Project Management — at the University of Limerick). Other universities offering masters' courses include Birmingham, Cranfield School of Management, Henley Management College, Heriot-Watt, Lancaster, Loughborough, Reading and UMIST (Manchester), and the subject of project management is included as an option in very many others.

Bibliography

The following lists all the references in this book, but in addition gives in most cases a guide to the nature and content of the reference. This is intended to help readers to judge which of the books listed are of particular relevance to them.

Ahuja H.N., Dozzi S.P. and AbouRizk S.M. *Project Management; Techniques in Planning and Controlling Construction Projects*. John Wiley, New York, 1994, 2nd edn. ISBN 0 471 59168 8

The material covered by Ahuja et al. is broadly similar to that contained in this present book (J.F. Woodward) with regard to the methods and techniques, but it goes further into these methods and explains many of them in mathematical terms. It is well illustrated by examples, and will present an opportunity for students and other readers to practice the calculation methods. There is less emphasis on the organisational aspects of project management such as procurement routes, people and systems.

Association for Project Management. *Body of Knowledge*. APM, High Wycombe, 1996.

This booklet is published by APM to outline the areas of knowledge which are relevant to the practice of project management. It is not a textbook encompassing the whole subject, but rather a listing of topics and the level of understanding required by practitioners, together with notes on the experience which is necessary. It is therefore a useful guide to those who seek to develop their careers in project management, and especially on routes to qualification.

Association for Project Management. *Standard Form of Project Management Agreement*. APM, High Wycombe, 1993.

This document sets out a standard form for formal terms of engagement between clients and independent project management specialist professionals.

Atkinson G. *A Guide Through Construction Quality Standards*. Van Nostrand Reinhold, Wokingham, 1987.

An early view of the rapidly evolving subject of quality assurance and particularly its associated certification schemes.

Atkinson G. *Construction Quality and Quality Standards, The European Perspective*. E & F N Spon, London, 1995. ISBN 0 419 18490 2

One of the few books devoted to quality in construction. About 20% sets out the special factors of construction, and the rest gives details of BS 5750, and its successor ISO 9000, and all the relevant published standards for Europe. Valuable reference book.

Baden Hellard R. *Total Quality in Construction Projects: Achieving Profitability with Customer Satisfaction*. Thomas Telford, London, 1993. ISBN 0 7277 1951 3

Most books on total quality management (TQM) relate to practice and experience in manufacturing industry, and only recently have there been any which are specifically concerned with project-based work such as construction. Constuction does present a special case for quality, and Hellard's book does cover this fully.

Baden Hellard R. *Project Partnering: Principle and Practice*. Thomas Telford, London, 1995. ISBN 0 7277 2043 0

The procedure of partnering embraces much of what is a fundamental long-term aim of project management, namely the achievement of a project by the cooperation of all parties to their collective satisfaction. It is based on the elimination of conflict, something which has bedevilled the industry for a long time. This book gives a good introduction to an approach which is still in its infancy.

Badiru A.B. *Project Management in Manufacturing and High Technology Operations*. Wiley-Interscience, 1996, 2nd edn. ISBN 0 471 12721 3

A very good general introduction to the principles of project management, covering also details of many of the techniques. Examples are quoted of major high technology projects, which make interesting reading. Of special relevance to readers in manufacuring industry.

Barnes M. *CESMM3 Handbook* [Civil Engineering Standard Method of Measurement]. Thomas Telford, London, 1992. ISBN 0 7277 1658 1

The Standard Method of Measurement prepared and published on the authority of The Institution of Civil Engineers is a central feature of most UK contracts in civil engineering for a number of years, and has been the subject of earlier editions of this book. It sets out in detail the way in which work should be measured and paid for, and is usually adopted within the contracts entered into by the major clients of heavy construction. The author has been deeply involved in the preparation of contract documentation for the institution, and gives a clear explanation of CESMM3, with many detailed examples.

Barrie D.S. and Paulson B.C. *Professional Construction Management*. McGraw-Hill, London, 1992. ISBN 0 0700 3889 9

The US practice of engaging professional consulatants to advise on the technology of construction methods is discussed in detail, along with the management of such a means of procurement.

Bennett J. *International Construction Project Management*. Butterworth Heinemann, Oxford, 1991. ISBN 0 7506 1330 0

Devoted to the specific subject of international projects it covers the area well and makes useful comparisons between UK, US and Japanese practices and terminology. It includes consideration of systems, quality, risk, control, and fast-track projects.

Bentley C. *Computer Project Management*. Wiley, 1982. ISBN 0 471 26208 0

Concentrates on the management of computer systems, but does not include much on project management concepts, and is not about the use of computers in project management. Of main interest to those in IT jobs.

Bergen S.A. *Project Management: An Introduction to Issues in Industrial Research and Development*. Basil Blackwell, Oxford, 1986. ISBN 0 631 14705 5

A particular case of project management is presented by development projects, where the scope is not clearly defined in advance, and the project may be abandoned at any time. The book includes a section on the writing of corporate plans.

Briner W., Geddes M. and Hastings C. *Project Leadership*. Gower, Aldershot, 1996, 2nd edn. ISBN 0 566 07714 0

An organisational look is taken at project management, considering the ways in which teams operate in different types of project, and how individuals behave. The book identifies different types of project as 'concrete', 'temporary' or 'open', and discusses how the project teams might be structured and managed. The role of leadership is discussed, and the ways in which the project leader or manager can interact with the client and the other members of the project team. The book gives a very interesting approach which offers a new insight to project management, especially for engineers and others with a largely technical background.

British Property Federation. *Manual of the BPF System*. BPF, London, 1983. ISBN 0 900101 08 3

This document explains in detail both the philosophy and content of the BPF (the federation of building development companies) system for building design and construction; it is accompanied by a BPF Standard Form of Building Agreement.

British Standards Institution. *BS 6079 Guide to Project Management*. BSI, London, 1996.

This standard has been issued to establish recognised guidelines for the conduct of projects, especially major government-sponsored projects. It follows earlier standards BS 4335 and 6046 which relate specifically to network techniques.

265

British Standards Institution. *Handbook 22*. BSI, London.

Part 1 is concerned with quality assurance and Part 2 with reliability and maintainability, bringing together the relevant BSI and ISO standards.

Clamp H. *The Shorter Forms of Building Contract*. Granada, London, 1984. ISBN 0 246 11964 0

The material of the book concentrates on forms of contract for minor works, especially the JCT 80 Agreement for Minor Building Works.

Cleland D.I. and Kerzner H. *A Project Management Dictionary of Terms*. Van Nostrand Reinhold, New York, 1985. ISBN 0 442 21690 4

This dictionary defines succinctly the terms used in project management, and adds an explanation of many of the acronyms. There is no descriptive text or discussion in its 287 pages, but it is a comprehensive reference document.

Cleland D.I. and King W.R. *Project Management Handbook*. Van Nostrand Reinhold, London, 1988. ISBN 0 442 22114 2

A well-known text which covers the subject area with some theoretical depth.

Construction Industry Council. *Project Management Skills*. CIC, London, 1996. ISBN 1 898 671 06 0

A concise fourteen-page document, published by a task force (which had a wide range of professional organisations as members), listing the skills required by constructional professionals who are involved in the management of projects. However, those aspects of project management which are solely of interest to contractors and suppliers are not included.

Conti T. *Building Total Quality*. Chapman and Hall, London, 1993. ISBN 0 412 49780 8

In spite of the title, this book is not about quality in the building industry, but about the 'building' of a quality system in any organisation. It is based on the wide European experience of its author, and is a translation from the original Italian. It gives good coverage in the growth of interest in total quality management world-wide, but makes no specific reference to quality in project management.

Cooke B. and Jepson W.B. *Cost and Financial Control for Construction Firms*. Macmillan, London, 1979. ISBN 0 333 24096 0

Management accounting for construction contractors is the theme of this book, concentrating on costs, cash flow, payments, and the potential of management information systems. There is also an introduction to financial management for contractors.

Craig S. and Jassim H. *People and Project Management for IT*. McGraw-Hill, London, 1995. ISBN 0 07 707884 5

Based on practical experience of information technology projects, this book is relevant to IT managers who may not be familiar with project management. Several examples are included, with checklists, and emphasis on the people aspect.

Cushman R.F., Stover A.B., Sneed W.R. and Palmer W.J. *The McGraw-Hill Construction Management Form Book*. McGraw-Hill, New York, 1983. ISBN 0 07 014995 X

The phrase 'construction management' in this title refers to the US where it means a particular form of contract, where a professional construction manager independent of the client or contractor is engaged on a fee basis to manage design and construction. The book is based around this concept and describes in detail some of the forms of contract between client and construction manager. UK practice is somewhat different.

Davis, Belfield and Everest. *Spon's Civil Engineering Price Book*. E & F N Spon, London, 1984. ISBN 0 419 11990 6

This is a standard reference text for contractors' estimators, setting out information on unit costs, overheads, general items, price indices, dayworks, fees and extracts from relevant documents such as working rule agreements. It is a companion volume to the annually published *Architects' and Builders' Price Book*, and is edited by the same Chartered Quantity Surveyors.

Donaldson H. *A Guide to the Successful Management of Computer Projects*. Associated Business Press, London, 1978.

A good general introduction to the application of project management to computer projects, but not really applicable to other industries. The general style is very readable.

Edwards L. *Practical Risk Management in the Construction Industry*. Thomas Telford, London, 1995. ISBN 07277 1656 5

The subject of risk management has evolved in recent years from the manufacturing industries where the emphasis is to some extent concentrated on commercial risks related to markets. The case of construction is different and this is one of the first books to tackle that area specifically; it deals with risks to clients, contractors and professional consultants.

European Foundation for the Improvement of Living and Working Conditions. *From Drawing Board to Building Site*. HMSO, London, 1991. ISBN 0 11 701576 8

A short book bringing together six European studies, it gives a good summary of the way that good quality at the design stage is reflected later in building projects.

Fangel M. Twelve Methods of Project Start-up. *World Congress on Project Management*, Rotterdam. International Project Management Association, 1985.

Frame J.D. *Managing Projects in Organizations*. Jossey-Bass, San Francisco, 1987. ISBN 1 55542 031 1

This book is mostly devoted to management *by* projects, i.e. the use of project management methods to manage change within existing organisations. In doing this it deals with software and manufacturing, mostly in general terms without the complication of detailed techniques.

Frankel E.G. *Project Management in Engineering Services and Development*. Butterworths, London, 1990. ISBN 0 408 03957 4

A theoretical study of networks, with special reference to dealing with uncertainty and project appraisal. Based on postgraduate work at Massachusetts Institute of Technology. Considerable attention is given to development projects, with much emphasis on project appraisal, including utility theory, multi-attribute analysis and cost/benefit analysis.

Gilbreath R.D. *How to make business projects succeed*. Wiley, USA, 1986. ISBN 0 471-83910 8

The emphasis here is on business projects which are 'exciting and challenging' because they take place outside the scope of on-going mainline operations. They are temporary and goal-directed, require confrontation with the unknown and hence are full of risks.

Gildersleeve T.R. *Data Processing Project Management*. Van Nostrand Reinhold, New York, 1985, 2nd edn. ISBN 0 442 22851 1

This book contains a good introduction to the concepts of project management, and then illustrates them with reference to the development and management of data processing projects.

Hamburg M. *Basic Statistics*. Harcourt Brace Jovanovich, New York, 1974. ISBN 0 15 505105 9

There are many books on statistics, and their level of approach varies from the simplistic to the complex. Hamburg's book is at the right level to be understood by engineers and other professionals who are numerate, and does not require a prior knowledge of statistics to understand it. The concepts of variability and probability are well explained, as is their relevance to control.

Hamilton A. *An Introduction to Project Management*. Centre for Project Management, Limerick, 1995.

A well set out series of lectures introducing a range of concepts of project management.

Harris F. and McCaffer R. *Modern Construction Management*. Blackwell Scientific, Oxford, 1995, 4th edn. ISBN 0 6320 3897 7

The emphasis of this book is on the techniques used by contractors in construction management. As such it includes many aspects of project management, and deals with them in a practical and non-mathematical way. In addition it covers aspects of construction technology and management, such as plant selection, cost control, competitive bidding, cash flow management and financial planning; all of these are topics of specific interest to contractors. As a means of training it includes several simulation 'games' which are useful.

Hertz D. and Thomas H. *Practical Risk Analysis*. Wiley, Chichester, 1984. ISBN 0 4711 0144 3

This is a companion volume to *Risk Analysis and its Applications* (published by Wiley, Chichester, ISBN 0 4711 0145 1), and mostly consists of a series of cases of investment decisions, the area in which much of the early work in the subject was based.

Hill T. *Production/Operations Management*. Prentice Hall, London, 1991, 2nd edn. ISBN 0 13 723727 8

A UK view of the standard methods of production planning and control, based on considerable industrial experience.

House R.S. *The Human Side of Project Management*. Longman, Harlow, 1988. ISBN 0 201 12355 X

A simple book based on experience in USA using tables and cartoons, dealing mostly with behavioural issues. It includes one major and several minor case studies to illustrate themes.

ICE Conditions of Contract. There is a range of documents published by Thomas Telford Services Ltd on behalf of The Institution of Civil Engineers, covering several types of forms of contract including:
ICE Conditions of Contract 6th edn (1991), plus guidance notes; *The New Engineering Contract* (1993), ten separate documents; *Professional Services Contract* (1994), plus guidance notes; *The Adjudicator's Contract* (1994); *Federation of Civil Engineering Contractors' Sub-contract* (1973); *ICE Design and Construct Conditions of Contract*, (1992); *ICE Conditions of Contract for Minor Works* (1995).

Jackson M.J. *Computers in Construction Planning and Control*. Allen & Unwin, London, 1986. ISBN 0 046 24010 1

One of the difficulties with a book on this subject is that it is almost inevitably out of date as soon as it is published. It does however give a good background to the concept of using information systems in construction project management, including standard data processing in estimating, planning and control, dealing with uncertainty, and introducing simulation. It does not give a guide to commercial software packages.

Jain R.K. and Hutchings B.L. (eds) *Environmental Impact Analysis.* University of Illinois Press, Urbana, IL, 1978. ISBN 0 252 00696 8

While this book is now somewhat dated it was an early statement on the subject, and is a collection of conference papers covering a range of US experience in the fields of physical, economic and social impact.

Jessen S.A. *The Nature of Project Leadership.* Scandinavian University Press, Oslo, 1993. ISBN 82 00 21516 4

An interesting alternative perspective of the subject of project management, discussing how it differs from other forms of management. For readers who wish to gain a real insight into the behaviour of project organisations this is a most refreshing presentation.

Johnson R.A., Newell W.T., Vergin R.C. *Production and Operations Management.* Houghton Mifflin, Boston, 1974. ISBN 0 395 04695 5

This is a comprehensive and detailed guide to the classical quantitative techniques of operations and production management.

Joyce R. *The CDM Regulations Explained.* Thomas Telford, London, 1995. ISBN 0 7277 2034 1

A practical guide to the Construction (Design and Management) Regulations 1994 which relate to the legal responsibilities of all parties to construction projects with regard to the health and safety of people in the industry. The author is both a chartered civil engineer and a solicitor.

Karrass C.L. *Give and Take: Guide to Negotiating Strategies.* Crowell, New York, 1974. ISBN 0 690 00566 0

This is a practical guide to negotiation in any situation, not specific to projects. It does give a good insight into the thought processes which may at first seem unusual to technical people who are not familiar with behavioural approaches; some of the methods have been tested in classroom simulations of negotiation with considerable value.

Langford D., Hancock M.R. and Fellows R. *Human Resource Management in Construction.* Longman Scientific and Technical, Harlow, 1995. ISBN 0 5820 9033 4

What was at one time called personnel management or labour relations has now evolved into human resource management (HRM). As in many areas the construction industry is somewhat different from others, and this is set out by Langford et al.

Langford D. and Rowland V.R. *Managing Overseas Construction Contracting*. Thomas Telford, London, 1995. ISBN 0 7277 2029 5

International contracting has increased both in terms of its volume, and also in terms of the number of countries and companies competing for the work. This has given rise to project management problems which demand special consideration and attention. The authors give an overview of the subject and conclude with two major case studies.

Leech D. and Turner B. *Project Management for Profit*. Ellis Horwood, London, 1990. ISBN 0 137 21887 7

Mostly based on large projects in civil and mechanical engineering and the aircraft industry, this book is well presented and comprehensive, with the approach of a management scientist.

Lewin M.D. *Software Project Management, Step by Step*. 1988. ISBN 0 9627 0220 X

A detailed explanation of the approach to managing a software project.

Lichtenberg S. Project Management. *International Symposium of Project Management*. Sorrento, 1984.

This paper, presented by the President of the International Project Management Association, gives a clear insight into the problems of estimating the costs of major projects.

Loasby B.J. *Choice, Complexity and Ignorance*. Cambridge University Press, Cambridge, 1976. ISBN 0 521 21065 8

This book really has little to do with project management, although its title might be thought by many construction professionals to be a fair description of their industry. The reason for its inclusion here is the discussion it contains on cost/benefit analysis.

Lock D. *The Essentials of Project Management*. Gower, Aldershot, 1996. ISBN 0 5660 7745 0

Lock has developed his earlier book *Project Management* by extending it through successive editions to include commercial and accounting practices and engineering administration. His new book is now a condensed version of this concentrating on the esentials of the subject. It sets these out and illustrates them by simple examples, most of which are presented through the format of one specific software package.

Lock D. *Project Management*. Gower, Aldershot, 1996, 6th edn. ISBN 0 5660 7709 4

A standard textbook on the subject, now expanded over several editions; it has also been summarised in a simpler version, *The Essentials of Project Management*.

Love S.F. *Achieving Problem-free Project Management*. Wiley, 1989. ISBN 0 4716 3522 7

This is a book for a non-technical reader, using anecdotal cases to exemplify points; it is a good primer. Reference is made to down-sizing, an exercise which has to be managed as a project.

Manton D.J., Roebuck E.J. and Fordham, G.L. *Building and Extending a Radiology Department*. Royal Society of Medicine Services, London, 1991. ISBN 0 9059 5870 5

This is a very special and interesting case, dealing with all the administrative and decision processes; it would be an extremely useful reference for any project manager faced with a similar case. It goes into significant details about room layouts etc.

Marsh P.D.V. *Contracting for Engineering and Construction Projects*. Gower, Aldershot, 1984. ISBN 0 5660 2232 X

This is a general text about the processes of planning, tendering, and the operation of contracts in civil and heavy construction engineering rather than building. It does not examine in detail any one form of contract, but gives a good overview of contracting in a straightforward way. For someone not familiar with engineering contracting, and wanting help to decide which type of contact to use, this is a good text; for someone seeking answers on fine points of contract it does not go into great detail on specific individual contract forms.

Moder J.J., Phillips C.R. and Davis E.W. *Project Management with CPM, PERT and Precedence Diagramming*. Van Nostrand Reinhold, New York, 1983. ISBN 0 4422 5415 6

A thorough detailed description of CPM, PERT and precedence diagrams, making use of relevant statistical methods. It is one of the classical presentations of dealing with uncertainty in time estimates, and describes methods which are of special relevance to development projects which may be open-ended.

Morris P.W.G. *The Management of Projects*. Thomas Telford, London, 1994. ISBN 0 7277 1693 X

This is essentially a historical account of the development of project management, a summary of the discipline and its possible future. By describing several major projects it conveys something of the excitement of being involved in their management. The author has been chairman of the Association for Project Management for some years, and is well placed to describe the subject in the way that gives a good understanding of the practice of the discipline. As he says in the foreword it is not a 'how-to' book, but does mention some of the methods in their practical context. This book is an important contribution to the study of the subject of project management as a whole.

Morris P.W.G. and Hough G.H. *The Anatomy of Major Projects*. Wiley, Chichester, 1987. ISBN 0 4719 1551 3

Interesting case studies of a number of famous projects, including Concorde, Channel Tunnel, Advanced Passenger Train, Thames Barrier, Heysham Nuclear Power Station. Includes concepts of assessing success and failure, but does not go into detail on other management practices.

OECD. *Environmental Impact Assessment*. OECD, Paris, 1979. ISBN 92 64 11918 3

This small book sets out the European view of the subject, nation by nation, and thus shows the early work being done in the area.

Parris J. *The Standard Form of Building Contract, JCT 80*. Collins, London, 1985. ISBN 0 0038 3027 6

This book is well known as the definitive work on JCT 80 which for many years has been one of the most widely used forms of contract in the building industry, (as distinct from civil engineering). It gives a thorough grounding in its subject and includes a good reference section on cases and an index on clauses. It is therefore a very useful reference work.

Pilcher R. *Project Cost Control in Construction*. Blackwell Scientific, Oxford, 1994, 2nd edn. ISBN 0 6320 3637 0

This book looks at control of construction costs from the point of view of both the client of the construction process and the contractor, and discusses trade-offs between cost and programme.

Pilcher R. *Principles of Construction Management*. McGraw-Hill, London, 1992, 3rd edn. ISBN 0 0770 7236 7

The approach of this book is through the application of modern management methods to construction, including several of the quantitative methods. These include network planning, and also a look at optimisation and other operational research techniques. The book includes several exercises and their solutions.

Price S. *Managing Computer Projects*. Wiley, Chichester, 1986. ISBN 0 471 91113 5

A good simple book not saying much about project management as such, but giving a good technical description of computer projects.

Rakos J.J. *Software Project Management*. Prentice Hall, 1990. ISBN 0 13 826173 3

Devoted to the special case of project management in the software industry. Many of the techniques are familiar to all project managers, but the application is specific.

Reiss G. *Project Management Demystified*. E&FN Spon, London, 1992. ISBN 0 4191 69200 2

This is a fairly elementary text but one which gives a clear introduction to the basic concepts of project management. It draws a simple analogy 'A project is like herding a flock of sheep; some wander off and have to be brought back into line'.

Ronco W.C. and Ronco J.S. *Partnering Manual for Design and Construction.* McGraw-Hill, 1995. ISBN 0 0705 3669 4

A review of the develoment of US practice of construction procurement by partnering, whereby client and contractor jointly undertake a series of projects. Sample forms and agreements are also included.

Scherkenbach W.W. *The Deming Route to Quality and Productivity.* Mercury, London, 1982. ISBN 1 852 5108 2 X

The Deming route to quality management has in recent years become very popular in the manufacturing industry. As such some of the detail is not relevant to project management, but the underlying principles are important and are regarded as a classic approach.

Seeley I.H. *Building Economics.* Macmillan, Basingstoke, 1996, 4th edn. ISBN 0333 63835 2

This book gives a practical guide to controlling the cost of building projects at the design stage, rather than the on-site control of construction costs.

Silverman M. *A Short Course for Professionals.* Wiley, 1988. ISBN 0 471 61507 2

This is a self-learning text in A4 format, dealing with the standard network methods.

Smith N.J. (ed.) *Engineering Project Management.* Blackwell Science, Oxford, 1995. ISBN 0 6320 3924 8

This is a series of descriptions of large engineering projects, with the emphasis on costs.

Spinner M.P. *Improving Project Management Skills and Techniques.* Prentice Hall, 1988. ISBN 0 13 452 831 X

A discussion of the development of the methods of project management.

Spinner M.P. *Elements of Project Management.* Prentice Hall, Eaglewood Cliffs, 1992, 2nd edn. ISBN 0 1325 3246 8

A good basic text of the techniques of planning with several useful simple examples; it is useful for students who wish to become familiar with these methods. Little is said about the role of people, contracts or appraisal.

Stallworthy E.A. and Kharbanda O.P. *Total Project Management.* Gower, Aldershot, 1983. ISBN 0 5660 2427 6

This was one of the first comprehensive books on the subject and it covers most of the topics of project management. Significant attention is given to the characteristics of project managers since they 'have to get things done'. Leadership is discussed in terms of using the stick or the carrot, and the art of listening is emphasised. Some comparisons are made between the project management approaches of different countries.

Stallworthy E.A. and Kharbanda O.P. *International Construction and the Role of Project Management*. Gower, Aldershot, 1985. ISBN 0 5660 2546 9

A useful complement to these authors' technique-based books, this one describes the special problems encountered in international projects with different cultures, legal systems and infrastructure. It is well illustrated with several project cases, mostly of large projects, but unusually has much to be said about what can be learned from project disasters.

Stallworthy E.A. and Kharbanda O.P. *Project Teams; the Human Factor*. NCC Blackwell, Manchester, 1990. ISBN 0 8555 4013 4

This is a thorough study of the people involved in project management, both as individuals and as teams. There is thorough coverage of the characteristics required of project managers, and there is much of interest to anyone setting up a project team, especially for large projects. Leadership is discussed, and also how attitudes and approaches vary in different parts of the world.

Stevenson R.J. *Project Partnering for the Design and Construction Industry*. Wiley, 1996. ISBN 0 471 10716 6

This is a major US statement on the principles and practice of partnering, a method of procurement aimed at eliminating the confrontational earlier practices of the industry.

Thompson P. and Perry J. *Engineering Construction Risks*. Thomas Telford, London, 1992. ISBN 0 7277 1665 4

Risk management is an area which has special relevance in construction projects by virtue of their one-off nature, and this book gives and introductory overview of the subject.

Thompson P. *Organization and Economics of Construction*. McGraw-Hill, London, 1981. ISBN 0 07 084122 5

Now a little dated, but it gives a good description of the organisational background of construction contracts. It does include project management, but also construction technology.

Turner J.R. *The Handbook of Project-Based Management*. McGraw-Hill, London, 1993. ISBN 0 07 707656 7

A comprehensive and authoritative book which gives a sound analysis of the project approach to a wide range of situations in any industry. It covers the range from simple explanations of concepts to more complex theory, and is a very useful reference for all who are making a thorough study of project management.

Walker A. *Project Management in Construction*. Blackwell Science, Oxford, 1996, 3rd edn. ISBN 0 6320 4071 8

A systems approach to project management is the major theme of this book, and it gives an overall view of the subject, especially from the viewpoint of a chartered quantity surveyor. It thereby forms a bridge between modern management thinking and construction practice. There is some discussion of techniques, but not in great detail.

Wearne S.H. *Control of Engineering Projects*. Thomas Telford, London, 1989. ISBN 0 7277 1387 6

Related specifically to enginering this book outlines the planning and control of time, cost and labour with reference to several practical examples.

Woodward J.F. *Quantitative Methods in Construction Management and Design*. Macmillan, London, 1975. ISBN 0 3331 7720 7

A forerunner of this present book, it covers a range of applications of management science to the construction industry, including construction method study, linear and dynamic programming, optimisation and bidding strategy. It also contains some of the same basic material on arrow networks and gives several different practical cases. It says little however about people, contracts, costing or quality.

Woodward J.F. *Science in Industry, Science of Industry*. Aberdeen University Press, Aberdeen, 1982. ISBN 0 08 028451 3

An elementary introduction to the management of technology-based industry, including the language of industry, communication, quantitative methods, production, planning and control, design and forecasting.

Index

Learning
C